Lecture Notes in Physics

For information about Vols. 1–131, please contact your bookseller or Springer-Verlag.

Lecture Notes in Physics

Edited by H. Araki, Kyoto, J. Ehlers, München, K. Hepp, Zürich
R. Kippenhahn, München, H. A. Weidenmüller, Heidelberg
and J. Zittartz, Köln

213

Forward Electron Ejection in Ion Collisions

Proceedings of a Symposium Held at the
Physics Institute, University of Aarhus
Aarhus, Denmark, June 29–30, 1984

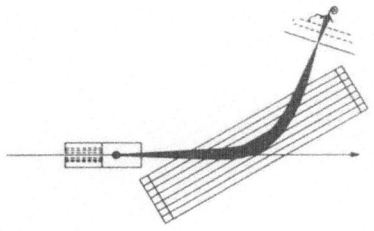

Edited by K. O. Groeneveld, W. Meckbach and I. A. Sellin

Springer-Verlag
Berlin Heidelberg GmbH 1984

Editors

K. O. Groeneveld
Institut für Kernphysik, Johann-Wolfgang-Goethe Universität, Frankfurt/M.
August-Euler-Straße 6, D-6000 Frankfurt/M. 90, FRG

W. Meckbach
Centro Atómico Bariloche, Comisión Nacional de Energia Atómica
8400 Bariloche, Argentina

I. A. Sellin
Department of Physics, University of Tennessee
Knoxville, TN 37916, USA

ISBN 978-3-540-13887-7   ISBN 978-3-540-39099-2 (eBook)
DOI 10.1007/978-3-540-39099-2

Originally published by Springer-Verlag Berlin Heidelberg New York Tokyo in 1984

2153/3140-543210

PREFACE

Since the coetaneous pioneering work of Crooks and Rudd at the University of Nebraska
and of Harrison and Lucas at the University of Sussex on electron ejection at forward
angles in ion-atom and ion-solid collisions, substantial and rapidly growing experimental and theoretical research concerning the physics of these processes has
occurred. Because of the large volume of the new experimental and theoretical results,
a two-day Symposium was held on related subjects hosted by the Physics Institute,
University of Aarhus, Denmark, June 29-39, 1984. The present volume contains the
Proceedings of this Symposium.

Because of his stature as one of the earliest investigators in the field, and his
continuing theoretical development thereof and interest therein, it seemed appropriate
to ask Joseph Macek of the University of Nebraska to deliver a keynote overview talk
on the background and current status of underlying theory. Though a conflict in
schedule prevented Professor Macek from favoring the Symposium with such a contribution, he offered instead to author a letter to be read at the Symposium summarizing
his views of the field and its prospectus for continued development. As many of the
sentiments expressed in his letter are shared by many other scientists, and indeed
provide a worthy rationale for working in the field, this foreword seems the appropriate place to quote Professor Macek's principal observations.

"... Your invitation ... (offers) ... the opportunity to say a few words on the
subject ... (of convoy electrons and related phenomena) ... a subject dear to my
heart. The growth of this field, both in quantity and quality, is truly astonishing.
The ingenuity of experiment is remarkable; new phenomena are uncovered so rapidly, it
is difficult to keep up. Your workshop is timely in providing an assessment of where
the field stands today. I only hope that somehow various contributions can be com-
piled so that those unfortunate enough to miss the occasion can benefit from your
effort to bring this field together.

"In looking back over developments, it seems to me that the chief concept underlying the whole field of convoy electrons and similar phenomena is the recognition
that there is really no sharp dividing line between reactions that produce bound
states and those that produce 'free' electrons; they are both part of the complete
picture of charge evolution in atomic collisions. It is heartening to see theorists
such as Bottcher, Shakeshaft, Devia, and Grün producing maps of dynamical charge
clouds to guide our thinking. On the experimental side, the complete maps produced
by Professor Meckbach and co-workers are remarkable in the detail that they uncover.
Above all, they show the rich structure that appears near the "forward peak".

"That such structure is worthy of our attention is attested to by the theory of Spruch and Shakeshaft who pointed out implications of the forward peak for second Born theories of electron capture. This work, and its experimental confirmation by Sellin, Groeneveld and others, could not have been possible without the acceptance of the commonality that underlies reactions involving bound as compared with continuum states. The work of Dahl and Rodbrø was instrumental in affirming the continuity of phenomena across ionization thresholds. It is now appropriate to review what has been accomplished and thereby set the stage for new exploitation of these ubiquitous "convoy" electrons.

"As a final remark let us recall the important work of one who is unfortunately no longer with us. I refer to our dear colleague, Klaus Dettmann. At the time of early observations by Harrison and Lucas in Sussex and Crooks and Rudd at Nebraska, the enthusiasm and excitement of Klaus was infectious. I am convinced that he saw more clearly than any of us at the time where these new discoveries would take us. He would certainly be gratified, as all of us early workers are, to see what the field has become today."

The organizers of the Symposium are particularly indebted to the staff of the Physics Institute, University of Aarhus, and especially to Helge Knudsen, for providing excellent facilities and many amenities which did much to enhance the experience of the Symposium attendees, numbering about 45. Both the venue and the menu proved ideal.

Multinational government agency support of the Symposium is also gratefully acknowledged. Principal agency support was provided by U.S. National Science Foundation, Division of International Programs and Physics Division; by FRG agencies BMFT-Bonn and DFG-Bonn; by the Institute of Physics, University of Aarhus; and by the Danish Physical Society. The co-operation of the publishers Springer-Verlag in arranging expeditious publication of the Proceedings is also gratefully acknowledged.

                    -- K.O. Groeneveld, W. Meckbach, and I.A. Sellin, Editors --

Contents

ELECTRON LOSS TO THE CONTINUUM FOR LIGHT IONS

M W Lucas, K F Man and W Steckelmacher
University of Sussex, Falmer
Brighton BN1 9QH, U.K.

Abstract

An attempt is made to examine the contribution that the study of "charge transfer into continuum states" and "projectile ionisation" has made to our overall understanding of charge exchange and ionisation and the link between them.
Velocity spectra for electrons ejected in the forward direction during collisions of H_2^+ and He^+ (0.8 - 2.8 MeV) with gas targets are shown together with the cross sections obtained by integrating such spectra over a range of velocities. The data are compared with the calculations of Briggs and Drepper and of Burgdorfer et al.

Introduction

The study of atomic collision processes involving charge transfer and ionisation, especially through the Coulomb field, has been a central part of physics since the discovery of radioactivity. Here in Denmark it is particularly appropriate that we should both acknowledge these early studies and assess the contribution of two recent processes to our overall understanding.

There are a number of excellent reviews and books discussing these topics in proper detail and giving a balanced account of the development of the subject but here I want to single out particular threads of thought from both classical and quantum mechanical approaches. In doing this I would note Bohr's "velocity matching criterion", which is still so much part of our intuitive thinking about charge exchange, and Born's celebrated approximation for dealing with the three body collision.

Historical Background

The classical approach to ionisation makes use of what we now call the "binary encounter approximation". Thomson (1912) calculated cross sections for the ionisation of a target atom by a bare projectile by assuming that the projectile interacts with only a single target electron; the nucleus and other electrons in the target playing no part except to provide a binding energy U of that electron. If the energy transferred to the target electron were ΔE and $\Delta E > U$, then ionisation would occur. The electron would be ejected with energy $E = \Delta E - U$. A weakness of

Thomson's model lies in its neglect of the fact that the struck electron
has an orbital velocity, this is important even when the projectile
energy E_p is high. The restriction was removed by Thomas (1927a) but
only to the extent that a single orbital velocity was chosen for the
electron. While this was a great improvement it does of course lead to a
sharp cut off to the ionisation cross section at the point where the
massive projectile transfers the maximum allowed energy to an electron
with kinetic energy E_2, i.e. $\Delta E = 4 \frac{m}{M} E_p + 4 \sqrt{\frac{m}{M}} E_p E_2$. Experimental
results do not show this cut off because, as the new wave mechanics was
beginning to show, there is a distribution of orbital velocities which
extends to infinity. The adverse effect of the cut off is shown in
figure 1 which I have taken from the paper by Rudd and Macek (1972).

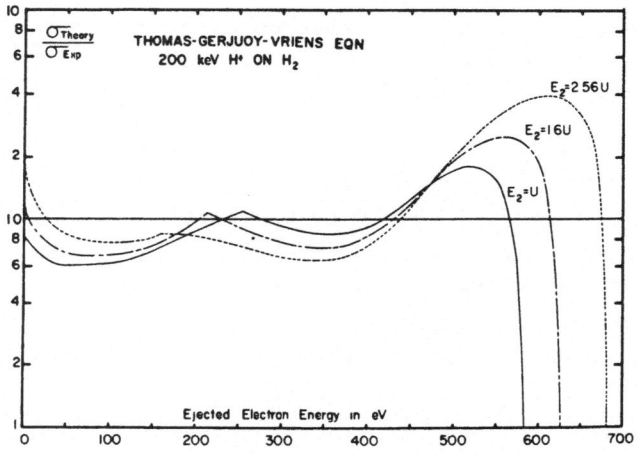

Figure 1

Ratio of theoretical to experimental cross
sections for ejection of electrons in
200 KeV $H^+ \rightarrow H_2$ collision. Theoretical results
calculated from Thomas - Gerjuoy - Vriens
equation using a delta-function distribution
of orbital velocities with energies as shown
in terms of the ionisation potential.
Experimental values from Rudd, Sautter and
Bailey (1966). From Rudd and Macek (1972).

It must have seemed that wave mechanics was able to offer a way out of
the difficulty and Thomas' work was largely forgotten. Even when
interest in classical theory was revived some thirty years later many of
Thomas' results were re-derived quite independently, but also more
generally, by Gerjuoy (1966) and by Vriens (1967). One must remember

that prior to the invention of the Cockroft-Walton and Van de Graaff generators, the only source of projectiles was radioactive decay so that earlier there was no great motivation to produce a generalised theory. Even in the 1960's there remained the difficulty with the cut-off but this could be removed by a slightly awkward marriage allowing classical projectiles to interact with Fock electrons.

My purpose in picking out Thomas' contribution to the problem of ionisation is not of course accidental in that he also had an interest in charge transfer which suffered a similar fate. As you know Thomas (1927b) envisioned that charge transfer required a double scattering mechanism whereby the incident projectile first struck the orbiting electron towards its host nucleus, from there it literally had to bounce off with the speed and direction of the original projectile in order to become bound to the fast projectile. In the first collision the target electron is scattered by the incident proton at an angle of 60° with respect to the incident proton. In the second collision the electron is re-scattered elastically by the target nucleus into the proton direction. Hence electron and projectile leave the collision together thereby facilitating capture. Conservation of energy and momentum require that the outgoing hydrogen atom appears at a small critical angle $\theta_T = \frac{\sqrt{3}}{2} \frac{m}{M} \simeq 10^{-3}$ rad. from the forward direction. Thomas also showed that the double collision process leads to a cross section which falls off very rapidly with increasing projectile velocity V_p. The model predicts V_p^{-11} but was later questioned since, as with the work on ionisation, a single velocity for the target electron was assumed rather that a distribution of orbital velocities which goes to infinity. Experiment was in no position to test these two predictions in the 1930's and again the new wave mechanics dominated subsequent theory.

The success of the wave mechanical treatments of ionisation and of charge exchange commencing with the work of Bethe (1930) and Brinkman & Kramers (1930) is famous. One might argue that the Bethe theory was the more immediately successful in that it correctly predicted the stopping cross section for fast particles in matter and as such was easier to test experimentally. In this sense it gave great credence to the Born approximation on which it was based. However, one must remember that such energy loss processes only require a knowledge of the total cross sections and not the doubly differential values (in energy and angle) that are a more sensitive test. It is interesting to note that Bethe did actually derive the velocity distribution of the ejected electrons and that suitable electron spectrometers existed to test these predictions (Hughes and Rojansky, 1929). However, possibly because attention

was moving towards nuclear physics, for which the stopping cross section
was of more interest, there seems to have been no thorough test of this
part of Bethe's paper, again for about thirty years. Interest in this
differential aspect was first awakened by the studies of Kuyatt and
Jorgenson (1963) and Rudd and co-workers, all at the University of
Nebraska. But one might also mention the first serious challenge to
the theory of the stopping cross-section which came with the discovery by
Barkas et al (1961) that the ranges of π^- mesons in matter is different
from that of π^+. A theory relating stopping to even powers of Z_p could
not accommodate this.

The charge exchange theory continued to develop but sufficiently
sensitive experimental tests of the new V_p^{-12} dependence for the cross
section at high velocities or the absence of a critical angle were not
available so that discussion was confined among the theorists. Shakeshaft
and Spruch (1979) remind us that this discussion was by no means
concerned with minor details and that Bohr and others viewed with
consternation the discrepancy between the Thomas (classical) V^{-11} and
the Brinkman-Kramers V^{-12} in a situation where the classical picture
should be just as correct as it is for Rutherford scattering. It is an
unhappy thought for an experimentalist that the first resolution of the
difficulty came solely from within the theory when Drisko (1955)
reasoned that since the classical double scattering is a two-step
process it should correspond to a second Born term. Drisko calculated
this second term and showed that indeed it does behave as V_p^{-11} for
asymptotically high velocities. Notice this only potentially resolves
the great discrepancy between the classical and quantum mechanical
approaches since there were still two important questions left un-
answered: "Does the second Born term dominate the first?" and, more
generally, "Does the Born series even converge? Might there not be an
even higher order term which dominates the second?". Partial and
affirmative answers to these essentially mathematical problems have been
given by Dettmann and Leibfried (1966, 1969), Corbett (1965) and
Shakeshaft and Spruch (1979) but only for a limited range of potentials
which do not actually include the Coulomb. Furthermore experimental
tests of the V_p^{-11} dependence seemed as far away as ever, particularly
when Briggs and Dettmann (1974) pointed out that at the very high
velocities of interest an alternative mechanism for charge exchange
would start to dominate. This is the process known as radiative electron
capture (Oppenheimer 1928) whereby the struck electron, instead of
scattering elastically off the target nucleus, adjusts its velocity to
be similar to that of the projectile when the collision region emits a

photon. The high velocity limit is only V_p^{-5} so that although the
numerical coefficient is smaller than that obtained by Drisko the
process will eventually dominate charge exchange through the Coulomb
field. Nevertheless all was not lost by the experimentalist when
Horsdal-Pedersen, Cocke and Stockli (1983) succeeded at long last in
demonstrating Thomas' critical angle phenomena for protons (v_p ~10 a.u)
incident on Helium. This is a much lower velocity than that at which
the V^{-11} dependence is expected to dominate the V_p^{-12} from the first
Born term (V_p~100 a.u. for Is-Is capture).

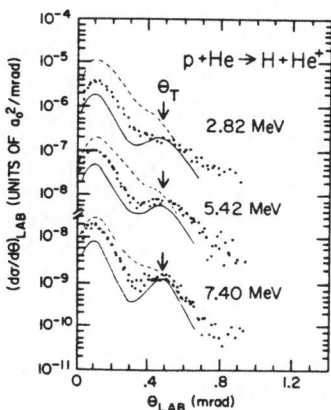

Figure 2

Differential cross sections for electron
capture to bound states at three projectile
energies. From Horsdal-Pedersen, Cocke and
Stockli (1983)

Before closing this selective review of our understanding of charge
exchange and ionisation I should like to stress a point made by Sellin.
The theories of the two processes had essentially proceeded independently.
To some extent this was so even in the early work of Thomas; it was
definitely so when the classical theories were re-introduced. The
quantum mechanical theories were also two distinct threads, possibly
because of the different way even the first Born approximation has to
be applied to each problem. The need for the second Born term to
describe charge exchange but not ionisation encouraged this separation.

The Place of Continuum-Electron Capture (CEC)

The sensitive test of theories of ionisation initiated in the 1960's at
Nebraska consisted of a series of experiments to study the doubly
differential cross section for electron ejection during proton bombard-

ment of light gases, particularly Helium. It soon became apparent that neither Binary Encounter nor Born approximation theories would reproduce the experimental results, particularly at small angles to the projectile beam, figure 3 (There was also a major discrepancy at angles near 180° but this is not central to the present discussion). The experimental data can also be displayed with electron energy as the abscissa and when this is done a hump appears in the 10° curve. This hump shifts with the beam energy but is not sensitive to the target gas. Oldham (1965, 1967) suggested this hump occurred because the ougtoing protons could "carry along" electrons from the target gas. These electrons would have to be moving with the protons but not bound to them.

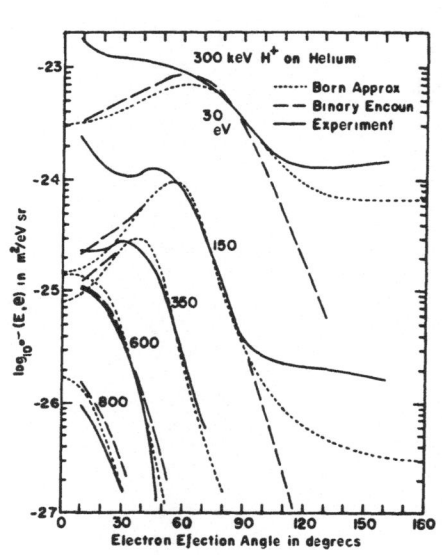

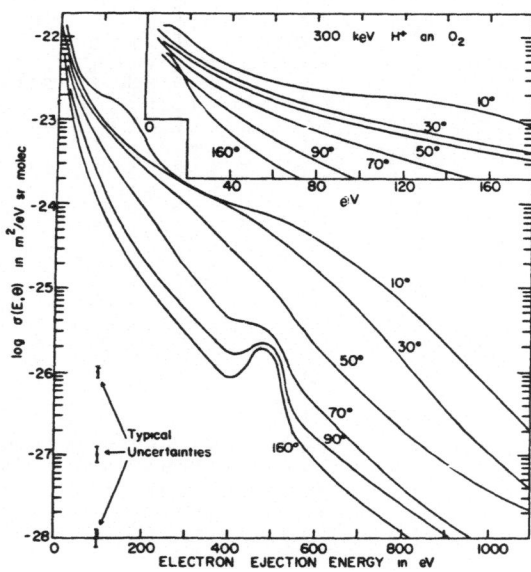

Figure 3

Comparison of doubly differential cross sections: from experiment (Rudd, Sautter, Bailey, 1966), from binary encounter calculations (Bonsen and Vriens, 1970) and from Born approximation calculations. From Rudd and Macek (1972)

Figure 4

Doubly differential cross sections for ejection of electrons from oxygen by 300 KeV protons. From Crooks and Rudd (1971).

Here, at last, was the connection between charge exchange and ionisation, first set down by Rudd and Macek (1972) and subsequently tested by Rodbro and Andersen (1979). Prior to this however, the question arose as to how to treat the problem theoretically; should one attempt to modify the theory of ionisation, an approach adopted by Salin (1969, 1972), or should one try to extend the theory of charge-exchange. Macek (1970) has used the latter approach, as have we, but some considerable difficulties are still to be overcome if these two mechanisms are to be properly combined to produce the complete electron ejection spectrum and avoid the double counting problem raised by Macek right at the beginning, yet such a cause of confusion in later work. It is to be emphasised that the distinction between the direct and exchange amplitudes used in the charge exchange approach is unclear in anything beyond first order and even in first order the cancellation provided by the "counter term" is only approximate except at high velocities. The failure of Duncan et al (1977) and of ourselves, to demonstrate the interference affects central to Macek's approach is cause for concern. This probably means a second order prescription is needed to obtain the whole secondary electron spectrum and this may influence the shape of the peak which, when observed at 0° rather than 10° is the hallmark of what was originally called "charge transfer into continuum states". As well as providing the link between our ideas about charge exchange and ionisation the study of CEC provided the direct experimental evidence of the need for second Born terms in charge exchange theory. This evidence significantly predates the successful observation of the critical Thomas angle θ_T.

Early first order theories based on the charge exchange approach suggested the peak would be symmetric in velocity space but Shakeshaft and Spruch (1978), reasoning that a charge exchange approach was the correct description, pointed out that the double collision mechanism would dominate at velocities as low as $V_p \sim 10$ a.u. provided that capture was into high $n\ell m$ states. In particular capture to high ℓ states would affect the symmetry of the continuum peak. Hence a study of the peak shape provides a sensitive test for second Born contributions which circumvents the difficulties inherent in both the search for the V_p^{-11} term, for charge transfer to bound states, and the very small critical angle scattering. Breinig et al (1980) have tested this idea using Ar^{18+} ($V_p = 18.1$ a.u.) incident on Helium and the results do confirm a contribution from second Born terms. Breinig et al (1982) also review other experimental evidence for which the asymmetry implies the necessity for the second Born terms, together with a counter proposal put forward by Chan and Eichler (1979). However another aspect of the behaviour of the cusp peak, namely that it shifts linearly rather than quadratically with electron ejection angle supports the view that the higher order terms are indeed important above about $V_p = 4.5$ a.u. (Macek et al 1981). However, this is not evidence that the double collision mechanism dominates at such low velocities, Macek (1983).

The second most important consequence of CEC then is the experimental support it has given to the theoretical belief in the

importance of second Born terms in charge exchange, but before leaving
CEC I would also like to draw attention to a remark originally made by
Crooks and Rudd (1970). The continuum capture, if viewed as a special
case of ionisation, must be incorporated in any theory of electronic
stopping. Since it has a cross section proportional to Z_p^3 it could
well have a bearing on the different stopping of the π^+ and π^- mesons.
At the moment this is still somewhat speculative since, in the
comprehensive studies of the Z_p dependence by Sellin and co-workers
using solid targets, the cubic term has not been observed. There are of
course other ways of obtaining the Z_p^3 term which are well known here at
Aarhus.

The Place of Projectile Ionisation (ELC)

When the projectiles carry electrons they may become ionised by
collision with the target atoms. The resulting electrons, ejected in
the moving frame of reference, also cause a peak in the forward spectrum.
This phenonemon was recognised by Rudd and Macek (1972) who showed a
peak produced by 300 KeV H_2^+ incident on He. It was also noted as a
possible contaminant to the signal we observed from solid targets
(Dettmann et al 1974) since, although protons were used, there was a
possibility that H^o might be formed by a charge exchange collision
within the target.

Now projectile ionisation bears a rather special relationship to
ionisation generally because of its ability to examine those electrons
having only very small energies relative to the projectiles from which
they were ejected. Suppose we set our spectrometer to examine electrons
moving with a small velocity $\pm \Delta v_{ep}$ in the projectile frame. Then in
the laboratory frame the electrons lie within the range $(v_p + \Delta v_{ep})$ to
$(v_p - \Delta v_{ep})$ i.e. $\Delta E_e = \frac{1}{2}m [(v_p+\Delta v_{ep})^2 - (v_p-\Delta v_{ep})^2]$
Now electron spectrometers with an energy resolution $\frac{\Delta E_e}{E_e} = 10^{-3}$ are not
uncommon so that close to the peak, where $E_e = \frac{m}{M} E_p$, we have
$$\Delta E_e \simeq 10^{-3} \frac{m}{M} E_p$$

combining equations 1 and 2 and remembering $\Delta E_{ep} = \frac{1}{2}m \Delta v_{ep}^2$

we obtain $$\Delta E_{ep} = \frac{10^{-6}}{16} \frac{m}{M} E_p$$

for the energy resolution in the projectile frame. For 1 MeV protons
this gives 3.4×10^{-5} eV, and, since the peak itself occurs at an

electron energy of some 545 eV it is well clear of the majority of the stray electron noise signal which occurs at tens of electron volts. Moreover the energy resolution in the projectile frame improves with increasing projectile mass (although use of more massive projectiles will eventually bring the signal into the more noisy regime unless the projectile energy E_p is increased). This ability to study electrons with very small energies relative to the projectile has been exploited in measurements of the shape of the loss peak by Meckback et al (1984) and in the detection of low energy autoionising levels of the projectiles (Lucas and Harrison 1972 , Suter et al 1979).

Menendez et al (1977) made a direct comparison of the peaks from He^{++} and He^+ incident on Argon finding them very similar, an entirely reasonable observation since the final state wave-function for both CEC and ELC may be taken as a Coulomb wave. Briggs and Drepper (1978) gave a theory for the electron loss process using this Coulomb wave. Since this is strictly an ionisation phenomenon the calculation was to first order and produced a velocity spectrum for the ejected electrons identical to the first order CEC theory; namely symmetric and of a width increasing with projectile velocity. Sellin and co-workers observed symmetric peaks for heavy ions incident on Argon but their width did not increase linearly with projectile velocity (Breinig et al 1981). Attempts to understand this have been made by Day (1980) and Briggs and Day (1980) but a more general picture showing how sensitive the results are to the initial state of the projectile has only been published recently (Burgdörfer et al 1983).

Now Briggs and Drepper also obtained cross sections for the loss process by integrating over a limited velocity range close to the peak. Most importantly and in contrast to those for continuum capture, these cross sections do not decrease strongly with increasing beam velocity a feature supported by Burgdörfer et al (1983). This is due to a simple kinematic effect and is well reproduced in the data, figure 5. However, the data of Strong and Lucas for He^+- He rise well above the theoretical curves at energies below about 2 MeV. Some rise is to be expected since the CEC process is not excluded from these data and Shakeshaft (1978) shows that for low energy projectiles CEC will be the dominant contribution to electron ejection. However, Shakeshaft's predicted maximum for CEC occurs at only 42KeV for protons on hydrogen and, while it would be a higher energy for He^+, 2 MeV is clearly unrealistic and indicates some erroneous signal was present in our earlier data. In repeating the measurements we have extended them to both H_2^+ and He^+ projectiles on to H_2, He, Ne and Ar targets using our

30° parallel plate analyser.

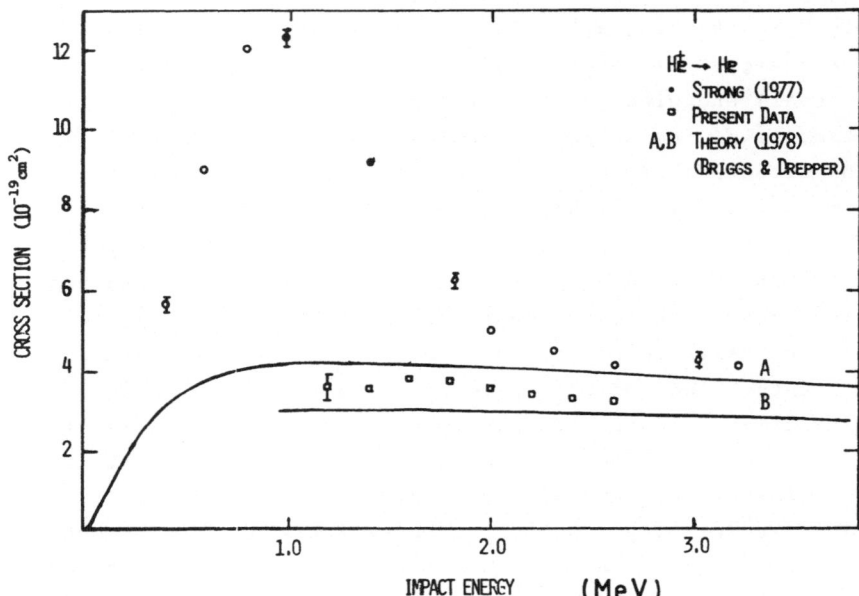

Figure 5

Cross section for electron loss in He$^+$ - He
collisions. Curves A and B are the theory of
Briggs and Drepper (1978) for ejection into a
solid angle 7.85 x 10^{-3} sr about the forward
direction. ϕ data of Strong and Lucas
(unpublished) ϕ present data
corrected for solid angle.

Because we have confined our measurements to an energy range well above
Shakeshaft's suggested 43 KeV we expect that the loss process is strongly
dominant, removing the need for a coincidence experiment. The data are
for ejection into a solid angle 1.13 x 10^{-3} sr about the forward direction and are obtained by integrating the differential curves over a
velocity range ± 0.25 a.u. either side of the peak. For this small
solid angle this velocity range incorporates almost all of the peak
above a smoothly drawn background. Because the ELC is a separately
identifiable process from that of target ionisation we believe it is
correct to subtract the background, figures 6 and 7.

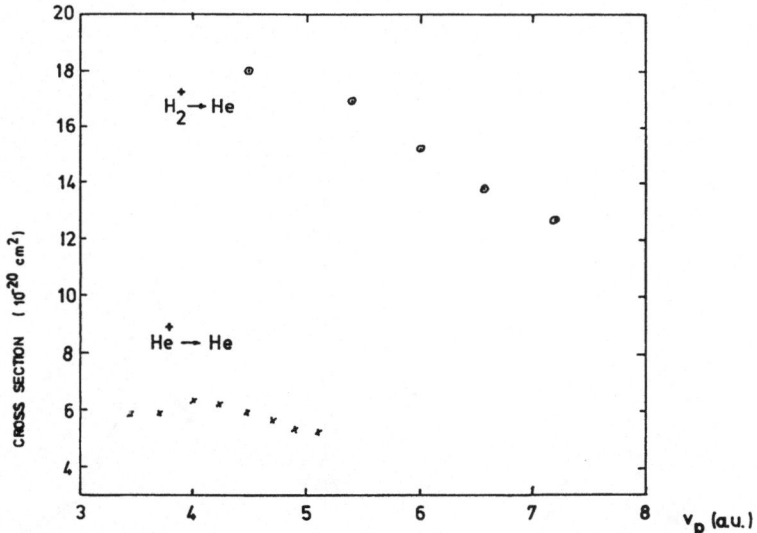

Figure 6

Cross sections for electron ejection using H_2^+ and He^+ projectiles incident on He. The ejection is into a solid angle 1.13×10^{-3} sr about the forward direction. Velocity spectra are integrated over $\pm$ 0.25 a.u. either side of the peak.

Some examples of the differential curves may be of interest, figure 7. These have a nearly constant full width at half maximum, Γ, which is not strongly dependent on velocity, e.g.:

$H_2^+ \rightarrow$ He (v_p= 4.4 - 7.2 au) Γ = 0.121 $\pm$ 0.018 au

$He^+ \rightarrow$ He (v_p= 3.4 - 5.1 au) Γ = 0.133 $\pm$ 0.020 au

If we scale our widths by the ratio of our acceptance angle (θ_o= 1.9 $\times 10^{-2}$ rad) to that of Breinig et al (1982) (θ_o= 3.14 $\times 10^{-2}$ rad) we can plot a figure 8 which is an extended version of the figure 11 given by Burgdörfer et al (1983). However, since the scaling for angle arises out of an assumption of isotropic emission (Γ = $\frac{3}{2} v_p \theta_o$), which is not obeyed in respect of velocity, this is illustrative only. (The data of Cranage et al (1982) for $H_2^+ \rightarrow H_2$ Γ = 0.65 $\pm$ 0.15, θ_o = 5 $\times 10^{-2}$ rad will not scale in this way.

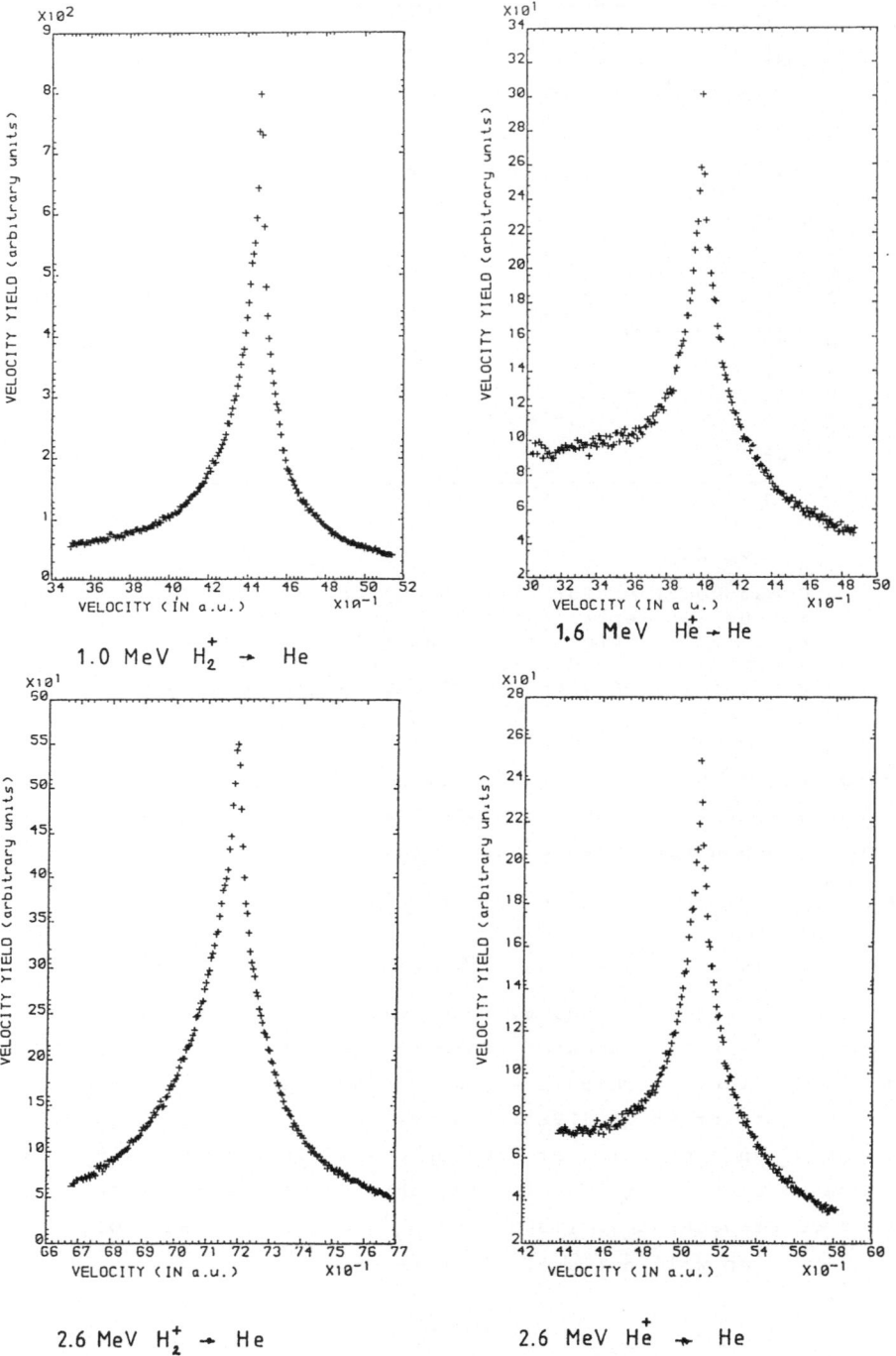

1.0 MeV $H_2^+ \rightarrow$ He

1.6 MeV $He^+ \rightarrow$ He

2.6 MeV $H_2^+ \rightarrow$ He

2.6 MeV $He^+ \rightarrow$ He

Figure 7

Differential velocity spectra for $H_2^+ \rightarrow$ He and $He^+ \rightarrow$ He, collecting angle 1.13 x 10^{-3} sr about the forward direction. Spectra are corrected for spectrometer pass band.

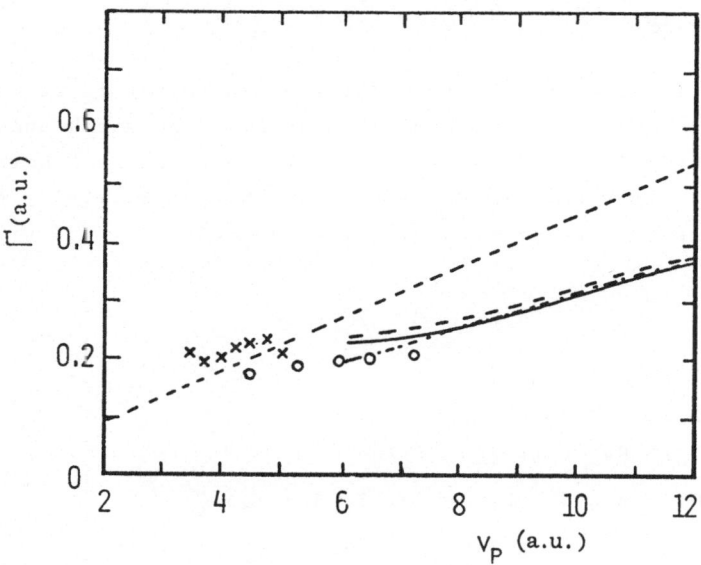

Figure 8

ELC cusp widths as a function of projectile velocity,
$Z_p = 2 : - \Gamma_{is}$ hydrogen target : $--- \Gamma_{is}$ helium target :
$-\cdot- \Gamma_{is}$ Argon target ($\theta_o = 3 \times 10^{-2}$ rad) After Burgdörfer
et al (1983). Dotted straight line is $\Gamma = \frac{3}{2} v_p \theta_o$
o $H_2^+ \to He$, X $He^+ \to He$ (scaled to $\theta_o = 3 \times 10^{-2}$ rad).

We have also examined the symmetry of these loss peaks and a first
analysis does suggest they are slightly wider on the low energy side.
This is most clearly seen in the data from H_2^+ projectiles where for
example $\frac{\Gamma_L}{\Gamma_R} \left(H_2^+ \to He \right) = 1.90 \pm 0.28$

in contrast to $\frac{\Gamma_L}{\Gamma_R} \left(He^+ \to He \right) = 1.08 \pm 0.15$

There is no clear indication that the asymmetry is velocity dependent
over the region examined so that a progressive failure of the Born
approximation at lower velocities is not seen.

With the same allowance for different solid angles we can plot
our total cross sections for the He^+ - He data on the Briggs and Drepper
curve, figure 5. We find rather satisfactory agreement but it is still
subject to reservations as to how to make a sensible correction for
different collecting solid angles now that we know the relation
$\Gamma = \frac{3}{2} v_p \theta_o$ is not obeyed.

ACKNOWLEDGEMENTS

I am greatly indebted to the late Klaus Dettmann for first interesting
me in charge transfer and to him, John Briggs, Joe Macek and
Robin Shakeshaft for guiding me through its mathematical horrors!
Ian Williams, Barry Farmery, Dave Hole and Norman Priestley have
given great assistance in obtaining the new data.

REFERENCES

Barkas W H, Dyer N J, Heckmann H H (1963)
          Phys Rev Lett $\underline{11}$, 26-28

Bethe H (1930)   Ann d. Physik. $\underline{5}$. 3. 325-400

Bonsen T F M and Vriens L (1970) Physica $\underline{47}$, 307-319

Breinig M, Elston S B, Sellin I A, Liljeby L, Thoe R, Vane C R,
Gould H, Marrus R and Laubert R (1980) Phys Rev Lett $\underline{45}$ 1689-1692

Breinig M, Schauer M M, Sellin I A, Elston S B, Vane C R, Thoe R S,
Suter M (1981) J Phys B$\underline{14}$, L291-L295

Breinig M, Elston S B, Huldt S, Liljeby L, Vane C R, Berry S D,
Glass G A, Schauer M, Sellin I A, Alton G D, Datz S, Overbury S,
Laubert R, Suter M (1982) Phys Rev A$\underline{25}$, 3015-3048

Briggs J S and Day M H (1980) J Phys B$\underline{13}$ 4797-4810

Briggs J S and Dettmann K (1974) Phys Rev Lett $\underline{33}$, 1123-1125

Briggs J S and Drepper   (1976) J Phys B$\underline{9}$ 2063
Briggs J S and Drepper F (1978) J Phys B$\underline{11}$ 4033-4039

Burgdörfer J, Breinig M, Elston S B, Sellin I A (1983)
          Phys Rev A$\underline{28}$ 3277-3290

Chan F T, Eichler J (1979) Phys Rev A$\underline{20}$, 367-8

Corbett J F (1968)   J. Math. Phys $\underline{9}$, 891-8

Crooks G B and Rudd M E (1970) Phys Rev Lett $\underline{25}$, 1599-1601
Crooks G B and Rudd M E (1971) Phys Rev A$\underline{3}$, 1628-34

Day M H (1980) J Phys B$\underline{13}$ L65-L68

Dettmann K, Harrison K G, Lucas M W (1974)
          J. Phys B $\underline{7}$ 269-287

Dettmann K and Leibfried G (1966) Phys Rev $\underline{148}$ 1271-3
Dettmann K and Leibfried G (1969) Z Phys $\underline{218}$ 1-24

Drisko R M (1955) Ph.D thesis, Carnegie Institute of Technology,
        unpublished

Duncan M M, Menendez M G, Eisele F L, Macek J
        (1977) Phys Rev A15 1785-6

Gerjuoy E (1966) Phys Rev 148, 54-59

Horsdal-Pedersen E, Cocke C L and Stockli M (1983)
        Phys Rev Lett 50, 1910-1913

Hughes A L and Rojansky V (1929) Phys Rev 34, 284-290

Kuyatt C E and Jorgensen T (1963) Phys Rev 131 666-675

Lucas M W and Harrison K G (1972) J Phys B5 L20-22

Macek J (1970) Phys Rev A1   235-241
Macek J (1983) Xlll   1 C P E A C p317-330
        (North Holland)
            eds: Eichler J, Hertel I V, Stolterfoht N
Macek J, Potter J E, Duncan M M, Menendez M G, Lucas M W,
        Steckelmacher W (1981) Phys Rev Lett, 46, 1571-1574

Mechbach W, Vidal R, Focke P, Nemirovsky I B, Gonzalex Lepera E
        (1984) Phys Rev Lett 52, 621-624

Menendez M G, Duncan M M, Eisele F L, Junker B R
        (1977) Phys Rev A15, 80-84

Oldham W J B (1965) Phys Rev 140, 1477-81
Oldham W J B (1967) Phys Rev 161, 1-6

Oppenheimer J R (1928) Phys Rev 31, 349-356

Rodbro M and Andersen F D (1979) J Phys B 12, 2883-2903

Rudd M E and Macek J (1972) Case Stud. At, Phys 3 47-136

Rudd M E, Sautter C A and Bailey C L (1966) Phys Rev 151, 20-27

Salin A (1969) J Phys B 2, 631-9, J Phys B2, 1225-6
Salin A (1972) J Phys B5, 979-86

Shakeshaft R (1978) Phys Rev A18, 1930-1934
Shakeshaft R and Spruch L (1973) Phys Rev A8, 206-215
Shakeshaft R and Spruch L (1979) Rev Mod. Phys 51, 369-405
Shakeshaft R and Spruch L (1978) Phys Rev Lett 41, 1037-1040

Suter M, Vane C R, Elston S B, Alton G D, Griffin P M, Thoe R S,
        Williams L, Sellin I A, Laubert R (1979) Z Phys A289
        433-434

Thomas L H (1927a) Proc. Camb. Phil. Soc 23, 713-716

Thomas L H (1927b) Proc. Roy. Soc. 114, 561-576

Thomson J J (1912) Phil. Mag. 23, 449-457

Vriens (1967) Proc. Phys. Soc. 90, 935-944

THEORETICAL DESCRIPTION OF THE CUSP ELECTRONS
EJECTED IN ASYMMETRIC HEAVY-ION COLLISIONS

D.H.Jakubaßa-Amundsen

Physik-Department, Technische Universität München, 8046 Garching
and GSI Darmstadt, Germany

Abstract

Starting from the Faddeev equations a series expansion for the transition amplitude for electron emission is given, which serves as a basis for the discussion of approximations used in the literature and their range of validity for a given collision system and momentum of the ejected electron. Both target and projectile electron emission will be considered. Emphasis is laid on the asymmetry of the forward peak and its variation with system parameters, such as collision velocity and charge ratio Z_P/Z_T. The theoretical results will be confirmed by a comparison with experimental data.

1.Introduction

Electron spectroscopy has received a great deal of attention lately, since the production of electrons by fast ion impact is of principal interest in many branches of physics. In particular, the so-called cusp electrons which are emitted with nearly zero velocity relative to the projectile, i.e. appear in the target frame in forward direction with a momentum that equals the collision velocity v have been focused in a variety of experimental and theoretical investigations (see i.e. Meckbach and Burgdörfer, proceedings of this conference). The spectral distribution of the cusp electrons provides a sensitive test for the applicability of first-order theories, and I will show in the following that these theories are invalid not only for electrons initially bound to the target, but also for electron loss from the projectile in asymmetric collision systems, even at very high impact velocity.

2.Faddeev Theory

The basic formalism for electron emission is most easily described in the three-particle picture, where the effect of all electrons of the projectile-target system except the active one is incorporated into the potentials. Let ψ_i^A be the initial electronic bound state (with A denoting either target T or projectile P), V_A the interaction field between the electron and A, and V_B the electronic perturbation in the initial channel. In the semiclassical theory the transition amplitude is given by (in atomic units)

$$a_{fi} = -i \int_{-\infty}^{\infty} dt \, \langle \Psi_f^- | V_B | \Psi_i^A \rangle \tag{2.1}$$

where Ψ_f^- is the exact solution to the electronic two-center problem with the boundary condition of a free electron in the final state. For fast collisions, which I shall concentrate on, the internuclear motion can be described by a classical straight-line path, thus neglecting the internuclear potential.

According to Faddeev[1], Ψ_f^- can be written as a sum of wavefunctions which are determined by a set of two coupled equations

$$|\Psi_f^-\rangle = |\Psi_{o,f}\rangle + |\Psi_1\rangle + |\Psi_2\rangle \tag{2.2}$$

$$|\Psi_1\rangle = G_B V_B |\Psi_{o,f}\rangle + G_B V_B |\Psi_2\rangle$$

$$|\Psi_2\rangle = G_A V_A |\Psi_{o,f}\rangle + G_A V_A |\Psi_1\rangle$$

where $\Psi_{o,f}$ is an electronic plane wave, and $G_{A,B}$ is a Green's function defined by $G_{A,B} = (i\partial/\partial t + \Delta/2 - V_{A,B} - i\varepsilon)^{-1}$ with $\varepsilon \rightarrow 0$. In order to avoid convergence problems from the long-range Coulomb fields it is assumed here and in the following that V_A, V_B are screened Coulomb potentials, with the well-defined limit of screening $\rightarrow$ 0 taken in the final T matrices. From (2.2), series expansions for Ψ_f^- can be derived. Using the definition of an eigenstate to nucleus A, $\Psi_f^A = (1 + G_A V_A) \Psi_{o,f}$ the series can be written in three different ways

$$|\Psi_f^-\rangle = (1 + G_B V_B + G_A V_A + G_B V_B G_A V_A + G_A V_A G_B V_B + \cdots) |\Psi_{o,f}\rangle \tag{2.3a}$$

$$= (1 + G_A V_A + G_B V_B G_A V_A + \cdots) |\Psi_f^B\rangle \tag{2.3b}$$

$$= (1 + G_B V_B + G_A V_A G_B V_B + \cdots) |\Psi_f^A\rangle \tag{2.3c}$$

where the form (2.3a) is a symmetric expansion allowing for both a projectile or target electronic final state.

For practical purposes, these series have to be truncated, such that it is no longer irrelevant, which of them is used. It is thus crucial to determine from the momentum k and the direction ϑ of the ejected electron relative to the parent nucleus, as well as from the charge ratio Z_A/Z_B, whether V_A or V_B will be the dominating potential, in order to truncate the appropriate series. An estimate for the potential acting on the electron a time τ after the collision can be found by using the Coulomb formula

$$V_A \approx \frac{Z_A}{k\tau} \quad , \quad V_B \approx \frac{Z_B}{\tau\left[(v - k\cos\vartheta)^2 + k^2\sin^2\vartheta\right]^{1/2}} \tag{2.4}$$

Restricting the discussion to the cusp electrons, one has to use (2.3c) for $k \to 0$ if $Z_A \gg Z_B$ (because V_A is dominant), and (2.3b) for $k \to v$ and $\vartheta \to 0$ if $Z_B \gg Z_A$ (as V_B is dominant), but in the cases $k \to 0$, $Z_A \ll Z_B$ as well as $k \to v$, $\vartheta \to 0$, $Z_B \ll Z_A$, the symmetric series (2.3a) should be preferred.

3. Approximations for the Calculation of the Forward Peak: Target Ionisation

Let me first consider the case where bare projectiles are impinging on the target atom, such that only target electrons are contributing to the forward peak. Solid state effects, as for example capture into high-lying bound projectile states which are eventually participating in the convoy electron production[2], will not be discussed here. In this case, A is identified with the target T, B with the projectile P.

a) First-order Born Approximation

The first Born approximation for target ionisation is obtained by retaining the first term in the series (2.3c), such that the transition matrix element reads

$$M_{fi}^{1.B.} = \langle \Psi_f^T \mid V_P \mid \Psi_i^T \rangle \tag{3.1}$$

Pioneer experiments on the forward peak in collisions of H^+ on He by Rudd and coworkers[3] have clearly demonstrated that (3.1) strongly underestimates the number of electrons ejected with beam velocity near 0°.

b) First-order Faddeev Approximation

From the estimate (2.4) it is obvious that in this case the projectile field must not be neglected in the expansion of Ψ_f^-. Using the symmetric series (2.3a) and retaining the first three terms $1 + G_P V_P + G_T V_T$, i.e.

$$M_{fi}^{1.FA} = \langle \Psi_f^P \mid V_P \mid \Psi_i^T \rangle + \langle \Psi_f^T \mid V_P \mid \Psi_i^T \rangle - \langle \Psi_{0,f} \mid V_P \mid \Psi_i^T \rangle \tag{3.2}$$

Macek[4] could explain the enhanced intensity of the cusp electrons. It originates from the normalisation of the projectile eigenstate Ψ_f^P, and diverges like $F_o = 2\pi Z_P/|\vec{k}_f - \vec{v}|$ as the momentum $\vec{k}_f$ of the electron approaches $\vec{v}$, unless the finite detector resolution is taken into account. However, a careful analysis of the peak shape[5] reveals an enhanced intensity of electrons with $k_f < v$ as compared to those with $k_f > v$, a feature which can not be reproduced by the symmetric factor F_o contained in (3.2).

c) Second-order Born Approximation

Incidentally, the insufficiency of a
first-order theory for the description of
charge transfer to the continuum (CTC)
which is the basic process leading to the
cusp electrons, is not surprising. It is
known from bound-state charge transfer
that the Brinkman-Kramers theory provides
the wrong high-velocity limit which has
only recently been verified experimentally[6], and also gives a wrong relative
occupation probability of the final
subshells. Therefore, Dettmann[7], Shake-

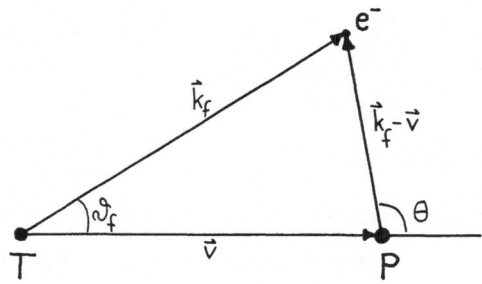

Fig.1 Momenta of the ejected electron
in the projectile or target frame

shaft[8] and coworkers have applied the second Born approximation, which consists in
retaining the fields up to first order in the series (2.3b):

$$M_{fi}^{2.B.} = \langle \psi_f^P | V_P + V_T G_0 V_P | \psi_i^T \rangle \qquad (3.3)$$

where G_o is the free propagator. The asymmetry property of the cusp electrons is
readily displayed by writing the doubly differential cross section in terms of a
partial wave expansion

$$\frac{d^2\sigma}{dE_f d\Omega_f} \sim \frac{1}{|\vec{k}_f - \vec{v}|} \sum_l a_l P_l(\cos\theta) \qquad (3.4)$$

where $E_f = k_f^2/2$, P_l a Legendre polynomial and θ the electron emission angle in the
projectile frame with respect to $\vec{v}$. It has been shown[9] that the assumption of continuity of the capture amplitude across the ionisation threshold leads to finite a_l
(including odd values of l) in the limit $\vec{k}_f \to \vec{v}$. This is satisfied for the second
Born theory[8,10], whereas a_l (l > 0) vanishes in the first-order theory. For zero
emission angle ϑ_f in the target frame, electrons with $k_f > v$ are emitted at $\theta = 0$,
such that $P_l = 1$, while electrons with $k_f < v$ appear in the backward direction
$\theta = \pi$, where $P_l = (-1)^l$, resulting in a discontinuity of the cross section at $k_f = v$.
The increased intensity at $\theta = 0$ may be explained by the fact that these electrons
are ejected in the same direction as the target motion (when viewed from the projectile frame).

Another distinction between the Brinkman-Kramers and the second Born approximation is found for the position E_{peak} of the forward peak as a function of emission angle ϑ_f. From (3.4) it can be verified[10] that the first-order theory shows a quadratic dependence on ϑ_f for small angles, while the second Born theory displays a linear decrease of E_{peak} with ϑ_f. The latter dependence is clearly verified by experiment in He^{++} on Ar (refs. 10,11) and H^+ on He collisions[12].

d) Impulse Approximation (IA)

If the projectile is much heavier than the target, a first-order treatment of the potential V_P in the T matrix is not sufficient. Instead, a consistent first-order expansion of Ψ_f^- in the weak target field V_T leads with the help of (2.3b) to

$$|\Psi_f^-\rangle = (1 + G_0 V_T + G_P V_P G_0 V_T)\,|\Psi_f^P\rangle$$

$$M_{fi}^{IA} = \langle\Psi_f^P|\,V_T(1 + G_P V_P)\,|\Psi_i^T\rangle \qquad (3.5)$$

where use has been made of the relations $G_P = G_0 + G_P V_P G_0$ and $\langle\Psi_f^P|V_P|\Psi_i^T\rangle = \langle\Psi_f^P|V_T|\Psi_i^T\rangle$. An insertion of a complete set of plane waves $\Psi_{0,\vec{q}}$ behind the operator $(1 + G_P V_P)$ in (3.5) allows for the description of the CTC process in terms of ejection of a target electron into an intermediate projectile state with momentum $\vec{q} + \vec{v}$, with a subsequent scattering by the target field into the final state Ψ_f^P.

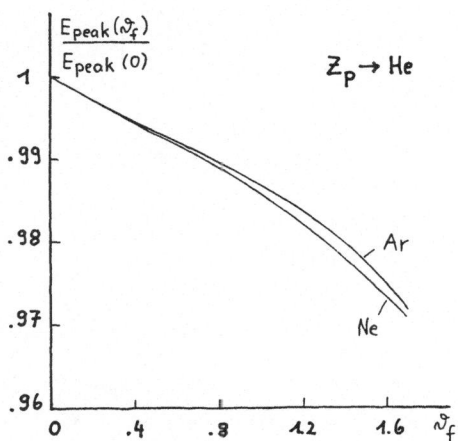

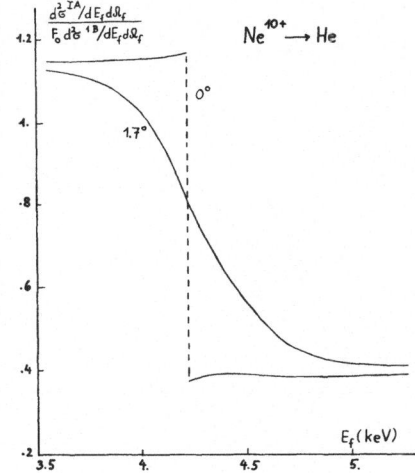

Fig.2 Ratio of peak energy to $v^2/2$ as a function of ϑ_f for collisions of Ar^{18+} ($v=18.1$ a.u.) and Ne^{10+} ($v=17.62$ a.u.) with He

Fig.3 Cross section ratio between impulse approximation and first Born theory times $F_0=2\pi Z_P/|\vec{k}_f-\vec{v}|$ at $\vartheta_f=0^0$ and 1.7^0 for 155 MeV Ne^{10+} on He

Off-shell effects for CTC are expected to be small[13] and neglected in the following.

As the impulse approximation reduces to the second Born theory for asymptotic collision velocities, it is expected to show similar features characteristic for a higher-order theory. Fig.2 displays the linear dependence of E_{peak} on ϑ_f for small angles in collisions of Ar^{18+} and Ne^{10+} with He at nearly equal impact velocities. In order to make the discontinuity of the differential cross section at $\vec{k}_f = \vec{v}$ visible, Fig.3 shows the ratio $d^2\sigma^{IA}/(F_0 d^2\sigma^{1.B.})$ which is finite at $E_f = v^2/2$. At nonzero angles the discontinuity is replaced by a rather steep fall-off of the ratio when E_f is increased.

In order to study the dependence of the discontinuity on the projectile charge and collision velocity, let me define the ratio

$$S = \frac{d^2\sigma/dE_f d\Omega_f(k_f^<, \vartheta_f = 0)}{d^2\sigma/dE_f d\Omega_f(k_f^>, \vartheta_f = 0)} \qquad (3.6)$$

where the electron momenta $k_f^<$ and $k_f^>$ are chosen such that $\eta = Z_p/|k_f - v|$ is equal (and $\gg 1$, but finite for numerical reasons) for both, with $k_f^< < v$ and $k_f^> > v$. As the electron is in a final projectile state, the field strength Z_p is taken as

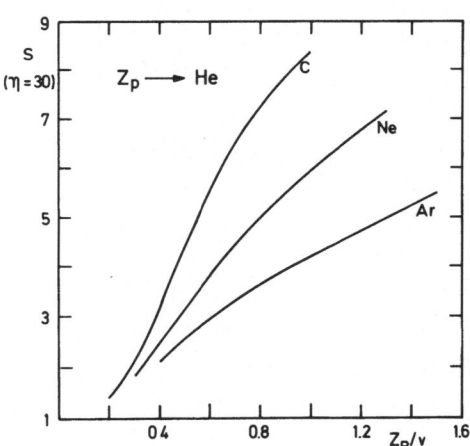

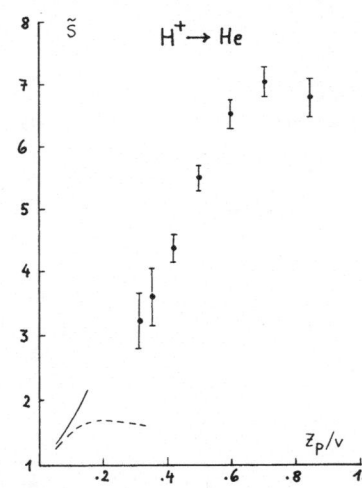

Fig.4 Discontinuity S at $\vartheta_f = 0$ and $\eta = 30$ as a function of inverse velo-city for collisions of C^{6+}, Ne^{10+} and Ar^{18+} with He

Fig.5 Discontinuity $\tilde{S}$ at $\vartheta_f \approx 0$ as a function of Z_p/v in $H^+ \to He$ col-lisions. Shown are results for the IA (——) and the asymptotic second Born theory (---). Data are from Dahl[12]

reference parameter also with respect to the velocity v. Fig.4 shows S in the IA for C^{6+}, Ne^{10+} and Ar^{18+} impinging on a He target. For large collision velocities, $v \gg Z_p$, a peaking approximation reveals[13] that in the limit of $k_f \to v$, $\vartheta_f \to 0$ the matrix element in (3.5) becomes proportional to a confluent hypergeometric function, the argument of which is $x \approx 16iZ_p/3v$ for $k_f \gtrless v$ and $x \approx 0$ for $k_f \lessgtr v$. Thus the discontinuity vanishes ($S \to 1$) for $Z_p/v \to 0$. At fixed Z_p/v which may also be considered as parameter of the initial perturbing field, S increases when Z_p/Z_T becomes smaller, i.e. when the relative influence of the target becomes stronger.

In the case of $H^+ \to$ He collisions, the systematics of the discontinuity has recently been investigated experimentally[12]. In Fig.5 the related quantity $\tilde{S} =$ $\overset{?}{k_f} F_o(\tilde{k}_f^?) \, d^2\sigma(\tilde{k}_f^?,0)/[\overset{<}{k_f} F_o(\tilde{k}_f^<) \, d^2\sigma(\overset{?}{k_f},0)]$ is shown, where $(\tilde{k}_f^< - v)^2 = (\overset{?}{k_f} - v)^2 = 2 \times 10^{-3}$ eV. However, the data are taken at relatively low v where the IA is no longer valid. Nevertheless, the trend shown in Fig.4 is in good agreement with the experiment. That the theory extrapolated to H^+ projectiles lies above the data results from the average over the angular acceptance in the experiment, which reduces the discontinuity (cf. Fig.3). For comparison, the second Born results are also shown for $H^+ \to$ He at $\vartheta_f = 0$, applying the high-velocity formula from ref.8. It is seen that even at rather high velocity, this formula breaks down, whereas at the highest velocity investigated, agreement is found with the IA within the numerical accuracy of the latter.

The discontinuity S reveals itself as a decisive parameter for the shape of the forward peak. Considering only partial waves with $l \leq 1$ in the expansion (3.4), the differential cross section may be approximated by

$$\frac{d^2\sigma}{dE_f d\Omega_f} = \frac{2\pi Z_p}{|\vec{k}_f - \vec{v}|} \, a \left[(S + 1) - (S - 1) \cos\theta \right] \qquad (3.7)$$

$$\cos\theta = \frac{k_f \cos\vartheta_f - v}{|\vec{k}_f - \vec{v}|}$$

which has the property that at $\vartheta_f = 0$, $d^2\sigma(k_f^<,0)/d^2\sigma(k_f^?,0) = S$, while $d^2\sigma$ is continuous for $\vartheta_f \neq 0$. In order to compare with experimental peak shapes, (3.7) has to be integrated over the angular acceptance ϑ_o, giving a simple analytic formula, and over the energy resolution ΔE_f. The constant a in (3.7) accounts for the absolute value which hitherto has not been determined experimentally. In the case of 155 MeV Ne^{10+} colliding with He ($\vartheta_o = 1.4°$, $\Delta E_f = 2.2\%$) the peak shape is compared with experimental data[14] in Fig.6. The dashed line is the formula (3.7) with S from Fig.4, and gives a reasonable fit to the peak shape. The agreement between the IA and (3.7) in the tails might be improved by introducing the energy dependence of the first Born theory into the constant a, as suggested from Fig.3. For the collision system $O^{8+} \to$

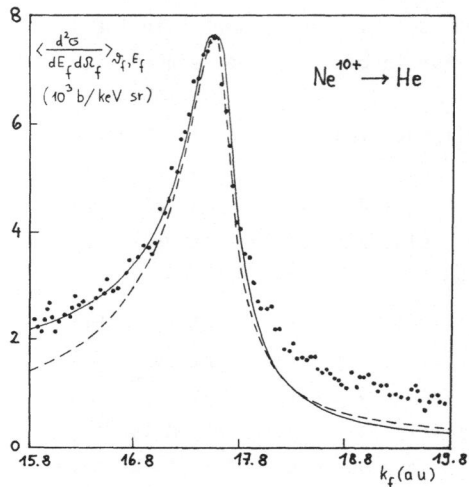

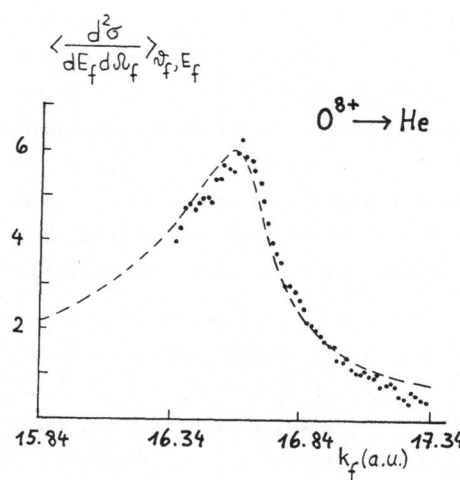

Fig.6 Differential cross section for
cusp electron emission in
$Ne^{10+} \longrightarrow$ He collisions at v=17.62
a.u. Full line, IA, dashed line,
(3.7) normalised at the peak.
Data are from Berry et al[14]

Fig.7 Differential cross section for
cusp electron emission in $0^{8+}\longrightarrow$
He collisions at v=16.64 a.u. The
dashed line is the formula (3.7),
the data are from Berry et al[14]
(arbitrary units)

He (ϑ_0 = 1.4°, ΔE_f = 1.4%) the formula (3.7) with S = 3.49 also compares well with
the experimental data[14]. Note that apart from the normalisation of the peak height
there are no free parameters in this theory.

4.Projectile Ionisation

When the projectile is not fully stripped, electron loss will in general provide
the dominant contribution to the forward peak at high collision velocities. Then in
the formulas (2.1) - (2.3), A has to be identified with the projectile P, B with the
target T.

a) First-order Born Approximation

The first Born approximation for projectile ionisation, which is obtained from
the first term in the series (2.3c), yielding

$$M_{fi}^{1.B.} = \langle \psi_f^P | v_T | \psi_i^P \rangle \tag{4.1}$$

has been frequently applied[15,16] for the description of the forward peak in systems
with $Z_P \gtrsim Z_T$ and large v. In a similar way as for target ionisation, the differen-

tial cross section near $\vec{k}_f = \vec{v}$ can be decomposed into partial waves

$$\frac{d^2\sigma}{dE_f d\Omega_f} \sim \frac{1}{|\vec{k}_f - \vec{v}|} \sum_{l=0}^{2n} a_l P_l(-\cos\theta) \qquad (4.2)$$

where in the first Born theory, l can only take even values[17,18] such that the cross section does not exhibit a discontinuity when θ is switched from 0° to 180^c. Experiments show[19] that the electron loss peak is much more symmetric than the corresponding CTC peak which would support the applicability of the first Born theory; however, the dependence of the peak width on velocity in collisions of e.g. Si with Ne and O with Ar show deviations[19] from first Born predictions, and a detailed study[20] of the angular distribution for H $\longrightarrow$ He collisions in the cusp region indicates that more partial waves than present in (4.2) should contribute.

b) Second-order Faddeev Approximation

For asymmetric systems with $Z_P \ll Z_T$ a treatment of the target field V_T in first order is not sufficient, as long as the collision velocity is not very much greater than Z_T. As in the cusp region both potentials will be important, the symmetric series (2.3a) for Ψ_f^- should be used. The second-order Faddeev approximation includes all terms explicitly written in (2.3a), such that the transition matrix element becomes

$$M_{fi}^{2.FA} = \langle \Psi_f^T | (1 + V_P G_P) V_T | \Psi_i^P \rangle$$

$$+ \langle \Psi_f^P | (1 + V_T G_T) V_T | \Psi_i^P \rangle$$

$$- \langle \Psi_f^T | V_T | \Psi_i^P \rangle - \langle \Psi_f^P | V_T | \Psi_i^P \rangle$$

$$+ \langle \Psi_{0,f} | V_T | \Psi_i^P \rangle \qquad (4.3)$$

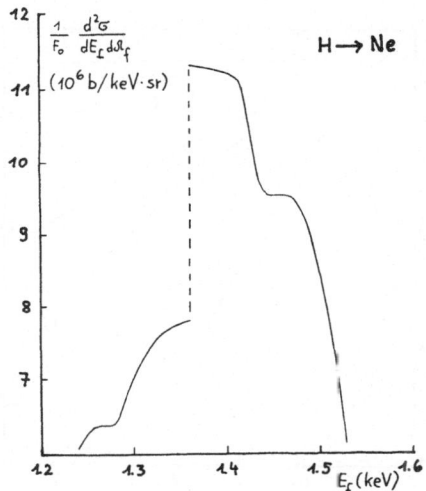

Fig.8 Cross section divided by F_o for electron emission in 2.5 MeV H $\longrightarrow$ Ne collisions at ϑ_f=0. Only the second term in (4.3) is included

where the last three terms correct for double counting and are nothing but the first-order Faddeev approximation. The first term, M_1, describes the charge transfer to the target continuum and influences the tails of the forward peak[21], whereas the second term, M_2, accounts for the projectile ionisation via an intermediate target continuum state and is dominant for $\vec{k}_f \approx \vec{v}$. At $\vec{\vartheta}_f = 0$, the matrix element corresponding to this second term[21] exhibits a phase proportional to $Z_p \cos\Theta$ which gives rise to a discontinuity as k_f traverses v. Unfortunately, this phase depends also on integration variables which cannot be fixed by a simple peaking approximation. In Fig.8 the discontinuity of the cross section calculated only from M_2 is clearly visible. In contrast to the target ionisation, the high-energy side is enhanced, because the electrons which are ejected with $\Theta = 0$ move in the same direction as the projectile from which they originate. Note that the wings of the peak will be influenced by M_1 which is not included in the calculation.

Again, one may define the ratio S according to (3.6) as a measure for the discontinuity. Fig.9 shows S as a function of Z_p/v for H on C, Ne and Ar targets. While the first Born theory always gives a ratio close to 1, the deviation from unity in the Faddeev approximation is the larger, the heavier the target atom and the lower the collision velocity, indicating the importance of the target field. In Fig.10 the absolute values of the differential cross section at $k_f \gtrless v$ and $\vec{\vartheta}_f = 0$ are compared to the first Born results. Considerable deviations are found even for a C target, and the Born theory becomes only valid at $v \gg Z_T$.

In conclusion, it has been shown for asymmetric collision systems, using the impulse approximation for target ionisation and the second Faddeev approximation for electron loss that the forward peak displays a discontinuity for zero emission angle

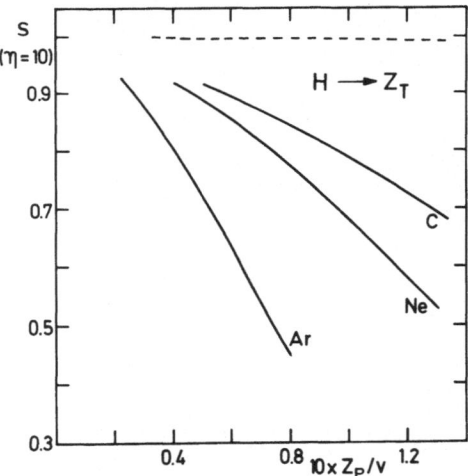

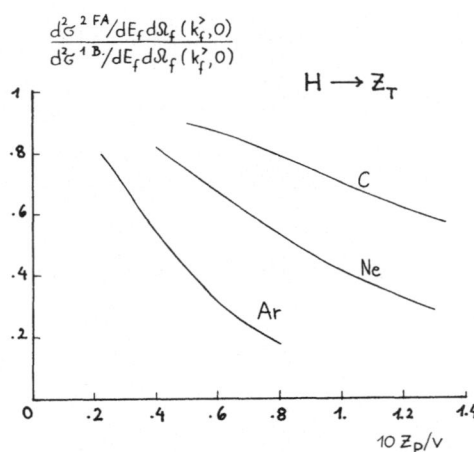

Fig.9 Discontinuity S at $\vec{\vartheta}_f=0$ and $\eta =10$ as function of inverse velocity for H colliding with C, Ne and Ar

Fig.10 Cross section ratio between the Faddeev and the Born approximation for H colliding with C, Ne and Ar at $\vec{\vartheta}_f=0$ and $\eta =10$

at k_f = v irrespective whether the electrons originate from the target or the projectile. The magnitude of this discontinuity can be closely related to the experimental peak shapes which provide strong evidence for the necessity of a higher-order theory.

References

1. L.D.Faddeev, Sov. Phys.- JETP 12, 1014 (1961)
2. H.-D.Betz, D.Röschenthaler and J.Rothermel, Phys. Rev. Lett. 50, 34 (1983)
3. M.E.Rudd, C.A.Sautter and C.L.Bailey, Phys. Rev. 151, 20 (1966)
4. J.Macek, Phys. Rev. A1, 235 (1970)
5. W.Meckbach, K.C.R.Chiu, H.H.Brongersma and J.W.McGowan, J.Phys. B10, 3255 (1977)
6. E.Horsdal Pedersen, C.L.Cocke and M.Stockli, Phys. Rev. Lett. 50, 1910 (1983)
7. K.Dettmann, K.G.Harrison and M.W.Lucas, J.Phys. B7, 269 (1974)
8. R.Shakeshaft and L.Spruch, Phys. Rev. Lett. 41, 1037 (1978)
9. M.W.Lucas, W.Steckelmacher, J.Macek and J.E.Potter, J.Phys. B13, 4833 (1980)
10. J.Macek, J.E.Potter, M.M.Duncan, M.G.Menendez, M.W.Lucas and W.Steckelmacher, Phys. Rev. Lett. 46, 1571 (1981)
11. R.O.Barrachina and W.Meckbach, Phys. Rev. Lett. 52, 1053 (1984)
12. P.Dahl, Contributed Paper to the XIII ICPEAC, Berlin, Book of Abstracts p. 374 (1983)
13. D.H.Jakubaßa-Amundsen, J.Phys. B16, 1767 (1983)
14. S.D.Berry, I.A.Sellin, K.O.Groeneveld, D.Hofmann, L.H.Andersen, M.Breinig, S.B.Elston, M.M.Schauer, N.Stolterfoht, H.Schmidt-Böcking, G.Nolte and G.Schiwietz, IEEE Trans. Nucl. Sci. NS30, 902 (1983)
15. F.Drepper and J.S.Briggs, J.Phys. B9, 2063 (1976)
16. M.H.Day, J.Phys. B13, L65 (1980)
17. J.Burgdörfer, M.Breinig, S.B.Elston and I.A.Sellin, Phys. Rev. A28, 3277 (1983)
18. H.Böckl, R.Spies, F.Bell and D.H.Jakubaßa-Amundsen, Phys. Rev. A29, 983 (1984)
19. M.Breinig, M.M.Schauer, I.A.Sellin, S.B.Elston, C.R.Vane, R.S.Thoe and M.Suter, J.Phys. B14, L291 (1981)
20. W.Meckbach, R.Vidal, P.Focke, I.B.Nemirovsky and E.Gonzales Lepera, Phys. Rev. Lett. 52, 621 (1984)
21. D.H.Jakubaßa-Amundsen, J.Phys. B14, 3139 (1981)

DOUBLE DIFFERENTIAL CROSS SECTION FOR ELECTRON
CAPTURE TO THE CONTINUUM WITH MOLECULAR PROJECTILES

C. E. Gonzalez Lepera* and V. H. Ponce

Centro Atomico Bariloche and Instituto Balseiro

8400 Bariloche, Argentina

It has been shown[1] that for H_2^+ in ident on thin carbon foils the yield per nucleon of convoy electron distributions is enhanced when compared to those observed with incident protons at the same velocity.

Using molecular continuum wave functions we obtained numerical results for the double differential cross section (DDCS) for electron capture from a 1s hydrogen-like initial state into the continuum of two correlated protons as a function of their interproton separation (R).

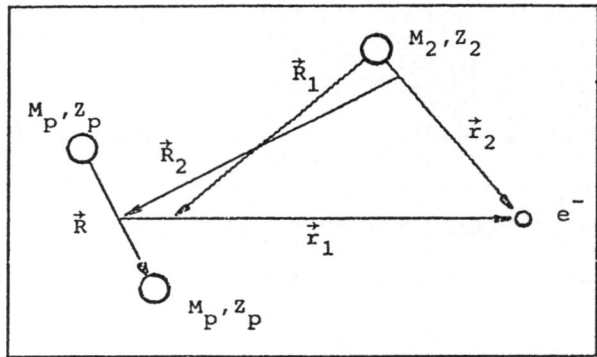

FIGURE 1: Coordinates system for the collision $2H^+ \rightarrow H$.

The system under study is represented in Fig. 1, where M_p, Z_p are the mass and charge of each proton respectively, and M_2, Z_2 are the mass and charge of the target nucleus. We assume that the distance R does not change during the collision (a.u. are used throughout this paper).

The first order Born approximation to the transition amplitude is

$$T_{if}^1 = \langle \chi_f | V_1 | \chi_i \rangle \tag{1}$$

where

$$\chi_i = e^{i\vec{K}_i \vec{R}_2} \Psi_{1s}(r_2) \tag{2} \qquad \chi_f = e^{i\vec{K}_f^1 \vec{R}_1} \Psi_{\vec{k}_1}(\vec{r}_1, \vec{R}) \tag{3}$$

are the initial and final states of the system respectively and

$$V_1(r_1) = - \frac{Z_p}{|\vec{r}_1 - \vec{R}/2|} - \frac{Z_p}{|\vec{r}_1 + \vec{R}/2|} \tag{4}$$

is the electron-projectile interaction potential.

The molecular continuum wave function $\Psi_{\vec{K}_1}(\vec{r}_1, \vec{R})$ has been calculated by Rankin and Thorson[2] and a detailed study of the problem can be found in the work of Ponomarev and Somov[3].

The integral expression for T_{if}^1 is rearranged as two terms, which are similar to those arising for the atomic projectile case. The dominant term is given by the electron wave function integrated over the coordinates of each proton $\Psi_{\vec{K}_1}^*(\pm\frac{\vec{R}}{2}, \vec{R})$ as can easily be demonstrated by expanding the wave function in powers of v and assuming that the projectile velocity is high ($v \gg 1$).

The triple differential cross section which is proportional to $|T_{if}^1|^2$ is then integrated over the allowed final projectile scattering directions to obtain the DDCS (Σ_1^2) through a suitable transformation into projectile transferred momentum $\vec{P} = \vec{K}_i - \vec{K}_f^2 = \vec{K}_i - \mu_1 \vec{K}_f^1 + \vec{K}_1$ where μ_1 is the reduced mass of the electron-projectile system. A further decomposition is made into P_1, P_t corresponding to the momentum transfer either parallel or normal to the velocity v.

In order to obtain numerical results we used the ground state of the hydrogen atom[4] as the target electron wave function.

The final expression for the DDCS is

$$\Sigma_1^2 = \frac{2^{10}\pi^2}{v^2} |\Psi_{E_k, L}(\frac{\vec{R}}{2}, \vec{R})|^2 \int_0^\infty dP_t \frac{P_t \, (1+(-1)^L \cos(RP_1\cos(\beta)))J_0(RP_t\sin\beta)}{(P_1^2+P_t^2)^2 \quad (1+P_1^2+P_t^2)^4} \tag{5}$$

where E_k is the electron energy in the projectile frame, L is angular momentum in the united atom limit (R=0) and β is the angle defined between the interproton vector R and the projectile velocity $\vec{v}$. In Eq. (5) we assumed $\kappa_1 \ll P$ which holds for capture to the continuum when $\kappa_1 \to 0$ and for high collision velocities $v \gg 1$.

The second-order Born term for the transition operator expanded in powers of the free-particle Green's operator G_0^+ is

$$T_{if}^2 = \langle x_f | V_2 G_0^+ V_1 | x_i \rangle \tag{6}$$

where $V_2(r_2) = -Z_2/r_2$ is the target nucleus-electron interaction potential.

Equation (6) is resolved in the Fourier space of the electron coordinates. A detailed study of the procedure has been given by Miraglia et al.[5] and therefore we will only show the final expression for the DDCS. That is

$$\Sigma_2^2 = \frac{2^{10}\pi^2}{v^2} |\Psi_{E,L}(\frac{\vec{R}}{2}, \vec{R})|^2 \int_0^\infty dP_t \frac{P_t[1+(-1)^L\cos(RP_1\cos\beta)J_0(RP_t\sin\beta)]}{(P_1^2+P_t^2)^2} \times$$

$$\times [\frac{1}{(1+P_1^2+P_t^2)^4} + \frac{1}{(P_1^2+P_t^2)^2(3P_1^2-P_t^2)^2} - \frac{2}{(P_1^2+P_t^2)(1+P_1^2+P_t^2)^2(3P_1^2-P_t^2)}] \tag{7}$$

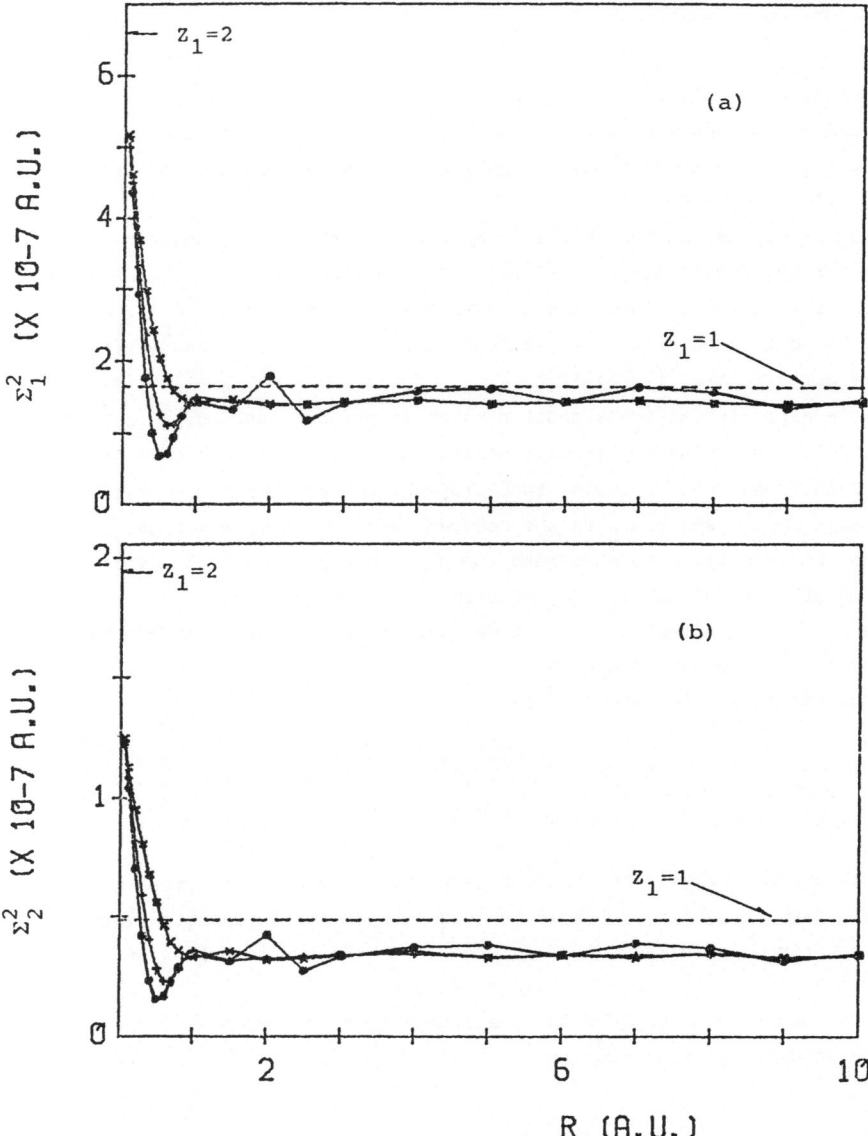

FIGURE 2: Double differential cross section (DDCS) as a function of interproton distance R for a collision velocity v =10 a.u. and an electron energy E=0.02 Ry in the projectile frame.

(a) Σ_1^2 , (b) Σ_2^2 ($\bullet$) $-\beta=0$. (+) $-\beta=\pi/4$. (x) $-\beta=\pi/2$.

The graphs shown in Fig. 2 correspond to the numerical integration of Eqs. (5) and (7) for Σ_1^2 and Σ_2^2 for a projectile velocity $v = 10$ a.u. and an electron energy $E = 0.02$ Ry in the projectile frame. The three curves represent different projectile orientations (β) relative to it's velocity. The limits $Z_1 = 1$ (dashed line) and $Z_1 = 2$ are those corresponding to the atomic and molecular (R=0) projectiles cases. A reasonable agreement is achieved with both limits.

A remarkable increase in the DDCS is seen for small R, even smaller than the measured interproton separation for the molecular ion H_2^+. But, at least for us, the unexpected results are the oscillations shown when the projectile orientation is coincident ($\beta = 0$) with the direction of motion. This fact suggests interference effects in the electron.capture process due to the presence of the second collision partner depending upon their relative orientation.

The same behavior of Σ_1^2 and Σ_2^2 has been observed for a higher collision velocity ($v = 20$ a.u.) showing a better agreement with the limits $Z_1 = 1$ and $Z_1 = 2$.

Results similar to those shown in Fig. 2 were also obtained keeping the molecular wave function $\Psi_{\vec{k}_1}(\vec{r}_1, \vec{R})$ inside the integral (Eqs. 5 and 7), as a further test to the approximation.

We wish to thank Drs. Rankin and Thorson for providing us with a copy of the Fortran 4 program for the calculations of the continuum molecular wave functions.

*Present address: Oak Ridge National Laboratory, Oak Ridge, TN 37831

References:

1) V. H. Ponce, C. E. Gonzalez Lepera, W. Meckbach and I. B. Nemirovsky, Phys. Rev. Lett. 47 572 (1981)
2) J. Rankin and W. R. Thorson, J. Comp. Phys. 32 437 (1979)
3) L. I. Ponomarev and L. N. Somov, J. Comp. Phys. 20 183 (1976)
4) V. H. Ponce, J. Phys. B 14 3463 (1981)
5) J. E. Miraglia and V. H. Ponce, J. Phys. B 13 1195 (1980)

DENSITY MATRIX DESCRIPTION OF COLLISIONAL ELECTRON
TRANSFER INTO THE CONTINUUM OF IONIC PROJECTILES

Joachim Burgdörfer
Institut für Atom- und Festkörperphysik, FU Berlin
W. Germany

Abstract

Recent theories for emission of cusp electrons resulting from either electron capture to continuum (ECC) or electron loss to continuum (ELC) show a large variety of anisotropies in the doubly-differential cross section (DDCS). We develop a unified description of cusp anisotropies in terms of the density matrix for low-lying continuum states. The anisotropy parameters β_k in the DDCS originating from partial-wave coherences can be expressed in terms of expectation values of the Runge-Lenz operator. Selection rules for vanishing β_k within the Born approximation are derived. Examples for the electron distribution following ECC and ELC will be discussed. The present approach provides also a unified description of anisotropic populations in both high Rydberg and low-lying continuum states.

I. Introduction

In 1970 Crooks and Rudd[1] discovered a cusp-shaped peak in the doubly-differential cross section (DDCS) for electron emission following ion-atom collisions close to the direction of the outgoing projectile. This electron cusp appears when the electron velocity vector in the laboratory frame, $\vec{v}_e$, approximately equals the projectile velocity, $\vec{v}_p$, in both magnitude and direction. In the same year Harrison and Lucas[2] observed a similar cusp in the electron distribution ejected by ions traversing solid foil targets. In both cases the cusp could be attributed to an excitation of similar final states in the low-lying continuum of the projectile with small velocities (in a.u.) $v \ll 1 (\vec{v} = \vec{v}_e - \vec{v}_p)$ as seen in the restframe of the projectile.

Since that time the process of electron transfer into the continuum has been investigated with increasing interest, both experimentally and theoretically. Several reviews[3-5] are now available which cover recent experimental developments in this field.

For ion-atom collisions under single-collision conditions two different mechanisms for the production of cusp electrons have been isolated: electron capture (ECC) and electron loss (ELC) to continuum. An ECC process requires a violent collision with a large momentum transfer to a target electron of the order $\cong v_p$ so that it finally ends up in a low-lying continuum state around the projectile. If, on the other hand, the projectiles carry one or more electrons into the collision an electron may be lost into the continuum during a soft collision with a small momentum transfer of the order $\cong$ ε_i/v_p. ε_i denotes the binding energy of the projectile electron to be ionized.

The mechanisms responsible for the production of "convoy" electrons in beam-foil interaction are much more involved. Recent experiments[5-7] seem to favour an interpretation in terms of an ELC process rather than an ECC process near the downstream surface of the foil but further investigations are necessary to unravel the underlying interaction processes.

A theoretical explanation for the occurrence of a cusp was first given by Macek and Rudd[8,9] who pointed out that ECC can be visualized as a smooth continuation across the ionization limit of capture into excited bound states with increasingly larger orbits such that the captured electron finally becomes unbound. Calculations for charge transfer from a hydrogenic 1s state to a projectile-centered continuum state are now available based on the first[8,10] and second[11,12] Born approximations, the impulse approximation[13] and a multiple-scattering theory[14]. Similar calculations for ELC have been performed for ionization of a hydrogenic 1s state[15] and arbitrary initial states (nlm)[16] using the first Born approximation.

Current theories for both ECC and ELC show a large variety of anisotropies and asymmetries to be present in the DDCS. A detailed analysis of nonisotropic structures has attracted considerable interest since recent experimental advances permit measurements of the three-dimensional electron distribution[17,18]. The Coulombic final-state interaction between the projectile ion and the ejected electron plays a decisive role in the anisotropic electron ejection. According to Wigner's threshold law[19] all partial waves are present in the zero-velocity ($v \to o$) continuum at threshold in the specific case of an attractive Coulomb field. The cusp anisotropy may therefore be analyzed in terms of the partial-wave population and of coherences between different partial waves. The goal of the present communication is to give a unified description of various cusp anisotropies in terms of the density matrix characterizing the collisional excitation of low-lying continuum states. We focus on (quasi) one-electron systems with a structureless continuum near threshold. Our approach is based on the dynamical 0(4) symmetry group of the Coulomb problem. It will be shown that a set of dynamical multipoles originally[20] introduced to classify bound-state coherences is well suited to describe also continuum-state coherences. The anisotropy parameters in the DDCS for electron emission can be expressed as expectation values of dynamical multipoles. By analyzing transformation properties with respect to parity and time reversal selection rules for anisotropy parameters within the Born approximation will be derived. Examples for ECC and ELC will be discussed. The present analysis suggests also a unified description of cusp electrons and collisionally excited Rydberg manifolds. Assuming the continuity of the multipole expectation values across the ionization limit the same parameters can describe the populations of both cusp and high-lying Rydberg manifolds. This approach is closely related to the extrapolation [21,22] of the total cross section across the ionization limit to verify the n^{-3} rule for bound-state capture.

2. Wave function at threshold

The origin of the cusp can be most easily understood by studying the zero-velocity limit ($v \to o$) of the Coulomb continuum wave function[23]. Using a partial-wave expansion of the wave function one finds in leading order of v,

$$\psi_{\vec{v}}^{\pm}(\vec{r}) = e^{\pm i\gamma} \sum_{1,m} (-1)^{\frac{1}{2}(1\mp 1)} \psi_{v1m}(\vec{r}) \quad Y_1^{m\star}(\hat{v}) \tag{2.1}$$

with

$$\psi_{v1m}(\vec{r}) = (\frac{2}{rv})^{1/2} \quad J_{21+1} (\sqrt{8Z_p r}) \quad Y_1^m(\hat{r}) \tag{2.2}$$

where $+(-)$ denotes the outgoing (incoming) Coulomb wave, J_k are the Bessel functions and Z_p being the charge of the projectile. γ denotes a divergent but ($1,m$)-independent Coulomb phase which is unimportant in the present context and will be omitted in the following. The important point is that all partial waves ($1,m$) exhibit a uniform $v^{-1/2}$ dependence of the amplitude in the limit of zero velocity in agreement with Wigner's threshold law[19] for an attractive Coulomb field. Recalling that the transition probability in the collision process depends on the square of the amplitude of the final state, the cross section is expected to diverge as v^{-1} in the limit of zero relative velocity between the electron and the projectile. This (integrable) singularity gives rise to a sharp peak ("cusp") in the electron distribution when $\vec{v} = \vec{v}_e - \vec{v}_p \to o$, or equivalently, $\vec{v}_e \simeq \vec{v}_p$ in the laboratory frame.

From a different point of view the $v^{-1/2}$ behavior of the amplitude (2.2) may be related to the density of bound states (Fig. 1). We expect the probability per unit energy interval (in a.u.) for finding an electron of a given ($1,m$) state at the position $\vec{r}$ to be constant as one crosses the ionization limit, i.e.

$$\lim_{v \to o} D(\epsilon > o) \mid \psi_{v1m}(\vec{r}) \mid^2 = \lim_{n \to \infty} D(\epsilon < o) \mid \psi_{n1m}(\vec{r}) \mid^2 \tag{2.3}$$

where

$$D(\epsilon < o) = n^3/Z_p^2$$
$$D(\epsilon > o) = v \tag{2.4}$$

are the densities of states per unit energy interval for bound and continuum states resp. Inserting (2.2) in (2.3) we recover the well-known Rydberg limit of the bound-state wave functions[23]:

$$\psi_{nlm}(\vec{r})_{n \to \infty} = (\frac{2Z_p^2}{n^3 r})^{1/2} \, J_{2l+1}(\sqrt{8\,Z_p r}) \, Y_l^m(\hat{r}) \tag{2.5}$$

Obviously the diverging normalization factor is closely connected with the density of bound states $\propto n^{-3}$ in an attractive Coulomb field. Whereas the occurrence of a cusp is simply a consequence of the Coulomb final-state interaction, the shape depends on the excitation amplitudes of individual partial waves and thus on the collision dynamics.

The dynamical O(4) symmetry of the Coulomb problem is related to a second constant of motion, the Runge-Lenz vector

$$\vec{A} \equiv \frac{1}{2} (\vec{p} \times \vec{L} - \vec{L} \times \vec{p}) - Z_p \frac{\vec{r}}{r} \tag{2.6}$$

($\vec{p}$: linear momentum) in addition to the angular momentum $\vec{L}$. $\vec{A}$ and $\vec{L}$ are the generators of the symmetry group O(4) that will be used to construct an appropriate multipole set parametrizing cusp coherences in the next section.

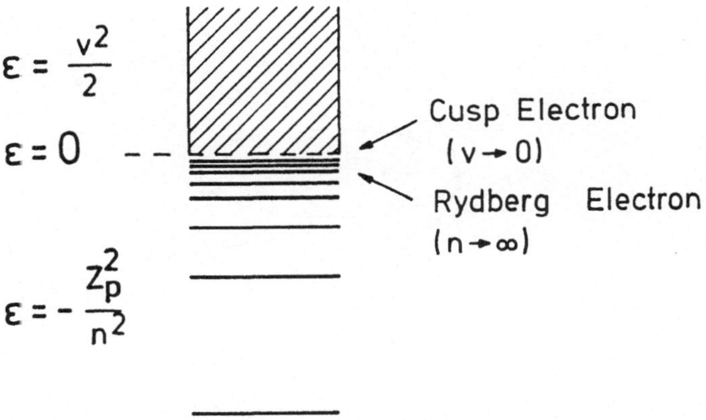

Fig. 1: Cusp and Rydberg states
in one-electron spectra

In a classical picture, $\vec{A}$ points from the nucleus to the "perihelion" of the electronic orbit and gives its principal axis (Fig. 2). $|\vec{A}|$ is proportional to the excentricity of the orbit.

Electronic states near the ionization limit correspond, classically, to parabolic orbits as the limit of both hyperbolas (continuum states) and ellipses (bound states). From Fig. 2 it is obvious that the Runge-Lenz vector is either parallel or antiparallel to the velocity of the cusp electron at large distances from the nucleus. In agreement with this classical picture one finds[16] $\psi_{\vec{v}}(\vec{r})$ to be eigenstates of the longitudinal component of $\vec{A}$,

$$A_v \, \psi_{\vec{v}}^{\pm}(\vec{r}) \underset{v \to 0}{=} \pm \, Z_p \, \psi_{\vec{v}}^{\pm}(\vec{r}) \tag{2.7}$$

with

$$\psi_{\vec{v}}^{\pm}(\vec{r}) = (\frac{Z_p}{4\pi^2 v})^{1/2} \, J_0 \, ([4Z_p r(1\mp\cos\theta)]^{1/2}) \tag{2.8}$$

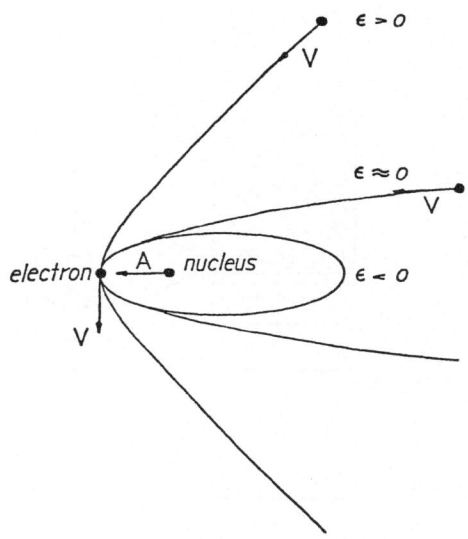

Fig. 2: Classical orbits for the Coulomb problem. States near threshold ($\varepsilon=0$) correspond to parabolas. At large distances from the nucleus $\vec{v}$ becomes (anti)-parallel to $\vec{A}$.

the zero-velocity limit of the continuum function in a parabolic representation and $\theta = \cos^{-1}(\hat{r} \cdot \hat{v})$. Similar to (2.5) also the parabolic representation possesses an analytic continuation immediately below the ionization limit in terms of the usual parabolic bound states with quantum numbers (n_1, n_2, m). The latter has been used in a group-theoretical calculation[16] of the transition form factor to cusp states determining ELC cusps in the Born approximation.

3. Dynamical multipoles for the DDCS

In a recent publication[20] a set of spherical multipoles has been introduced to classify the complete density matrix for a hydrogenic bound-state manifold with arbitrary principal quantum number n. Taking explicitly into account the $O(4)$ symmetry of the Coulomb field a complete set of operators can be constructed in terms of the two group generators, the angular momentum $\vec{L}$ and the normalized Runge-Lenz vector

$$\vec{a} = \frac{n}{Z_p} \vec{A} .$$
(3.1)

It can be shown[24] that the two pseudospins defined by

$$\vec{j}^{(1,2)} = \frac{1}{2} (\vec{L} \mp \vec{a})$$
(3.2)

obey the commutation rules of two independent angular momenta

$$[j_i^{(\alpha)}, j_k^{(\beta)}] = i \, \epsilon_{ikl} \, j_l^{(\alpha)} \, \delta_{\alpha,\beta} .$$
(3.3)

A set of multipole operators which will be named the dynamical multipoles in the following may be expressed in terms of tensor products of spherical multipoles $j^{(\alpha)}{}_q^k$ of the pseudospins. Restricting ourselves to the case of rotational symmetry with respect to the beam axis when the scattering angle of the deflected projectile remains undetected, only the spherical $q=o$ component is non-zero by geometrical symmetry considerations. The normalized dynamical multipoles are then given by[20]

$$U_{q=o}^k(\pi, k_1, k_2) = \frac{i^{k_1+k_2-k}}{2} \frac{\sqrt{(2k_1+1)(2k_2+1)}}{\langle j||j^{k_1}||j\rangle \, \langle j||j^{k_2}||j\rangle}$$

$$\cdot \sum_{q_1,q_2} \langle k_1 k_2 q_1 q_2 | ko \rangle (j^{(1)}{}_{q_1}^{k_1} \, j^{(2)}{}_{q_2}^{k_2} + (-1)^\pi \, j^{(2)}{}_{q_1}^{k_1} \, j^{(1)}{}_{q_2}^{k_2}) .$$
(3.4)

In (3.4) $\pi = o$ (1) denotes the even (odd) parity of the operator, $\langle j||j^{k_i}||j\rangle$ are the reduced matrix elements of the pseudospin tensors of rank k_i $(o \le k_i \le n-1)$ and the pseudospin quantum number j is related to the principal quantum number n by

$$n = 2j+1 .$$
(3.5)

Notice that in contrast to Ref. 20 the operators (3.4) are normalized in the sense of the trace metric in the Liouville formalism [25]. The dynamical multipoles possess a well-defined quantum number with respect to time reversal T. U_o^k is T-even (odd) if $k_1 + k_2 + \pi$ is even (odd).

The set of expectation values

$$\langle U_o^k(\pi,k_1,k_2)\rangle = Tr_n(\sigma\, U_o^k(\pi,k_1,k_2)) \tag{3.6}$$

with $\pi = 0,1$; $0 \leq k_{1,2} \leq n-1$ and $0 \leq k \leq 2n-2$ permits a parametrization of the complete n-shell density matrix σ. They can be expressed in terms of the spherical multipoles of the density matrix $\sigma_q^k(1,1')$ as

$$\langle U_o^k(\pi,k_1,k_2)\rangle = \frac{i^{k_1+k_2-k}}{2} \sum_{1,1'=0}^{n-1} \sqrt{(2k_1+1)(2k_2+1)(21+1)(21'+1)} \cdot \tag{3.7}$$

$$((-1)^{\pi}+(-1)^{1+1'}) \begin{Bmatrix} j & j & k_1 \\ j & j & k_2 \\ 1 & 1' & k \end{Bmatrix} \sigma_o^k(1,1')^*$$

with

$$\sigma_q^k(1,1') = \sum_{m,m'} (-1)^{1-m} \sqrt{2k+1} \begin{pmatrix} 1 & k & 1' \\ -m & q & m' \end{pmatrix} \langle n1m|\sigma|n1'm'\rangle . \tag{3.8}$$

The density matrix in the standard representation is given in terms of the transition amplitudes for the collision process, $t_{i\to n1m}(\vec{K})$, by

$$\langle n1m|\sigma|n1'm'\rangle = \frac{\delta_{m,m'}}{(2\pi)^2 v_p} \int d^3\vec{K}\, t_{i\to n1m}(\vec{K})\, t^*_{i\to n1'm'}(\vec{K}) \cdot \delta(\vec{K}\cdot\vec{v}_p+\Delta\varepsilon) . \tag{3.9}$$

The momentum transfer during the collision is denoted by $\vec{K}$, $\Delta\varepsilon$ being the change in the electronic energy. The density matrix is diagonal in m because the momentum transfer integral (3.9) extends over all scattering angles. This implies that only the q=o components of the spherical multipoles (3.8) are non-zero. The density matrix is, however, non-diagonal with respect to 1. Eq. (3.8), or equivalently (3.9), contains the maximum information on the collision amplitudes available in an angle-integrated experiment. The usual substate cross sections corresponding to diagonal elements in (3.9) do not completely specify the final-state population.

The dynamical multipoles contain the orientation and alignment parameters introduced by Fano and Macek[26] as a subset. An obvious advantage of the parametrization (Eq. (3.6)) is a decomposition of the density matrix according to their transformation properties with respect to parity (P) and time reversal (T): We find four different classes of parameters: the class P_eT_e even under both P and T, the class P_oT_e of parameters odd under P but even under T, the class P_eT_o of parameters odd under time reversal but even under P, and finally, the class P_oT_o odd under both P and T. Furthermore, the parametrization of the density matrix in terms of expectation values of $\vec{L}$ and $\vec{a}$ gives a direct insight into the physical meaning of various density matrix elements. Table 1 illustrates the relationship between operator expectation values and the density matrix for the n=2 level.

Class	k	operator	expectation value
P_eT_e	0	$U_0^0(0,0,0) = \frac{1}{2}$	$\frac{1}{2} Tr_{n=2}\, \sigma$
	0	$U_0^0(0,1,1) = \frac{1}{2\sqrt{3}}(L^2-a^2)$	$\frac{1}{2\sqrt{3}}(Tr_{2p}\sigma - 3<200|\sigma\,200>)$
	2	$U_0^2(0,1,1) = \frac{1}{2}(L_0^2 - a_0^2)$	$\frac{2}{\sqrt{6}}(<211|\sigma|211> - <210|\sigma|210>)$
P_0T_e	1	$U_0^1(1,1,0) = -\frac{1}{2}a_z$	$-Re <200|\sigma|210>$
P_0T_0	1	$U_0^1(1,1,1) = \frac{1}{2\sqrt{2}}(\vec{a}\times\vec{L}-\vec{L}\times\vec{a})_z$	$\sqrt{2}\, Im <200|\sigma|210>$

Table 1: Dynamical multipoles and density matrix for the n=2 level[20].
The only element of class P_eT_0 in n=2, $<L_y>$, vanishes due to rotational symmetry.

In the Rydberg limit (n >> 1) where the number of density matrix elements increases as $\propto n^3$ an orbital description in terms of a few multipoles $<U_0^k>$ is of particular interest since a complete determination of the density matrix seems to be impossible. The dynamical multipoles are supposed to sort out the most relevant information on the Rydberg orbit.

In order to investigate this conjecture we have calculated the Rydberg limit of Eq. (3.7). Using an approximate expression[27] for the 9j-symbol in the limit of large $j = (n-1)/2$ we have found to leading order in 1/n

$$<U_0^k(\pi,k_1,k_2)>_{n\to\infty} = \frac{i^{k_1+k_2-k}}{2n}(-1)^{k_2}\sqrt{(2k_1+1)(2k_2+1)}\begin{pmatrix}k_1 & k & k_2\\0 & 0 & 0\end{pmatrix} \qquad (3.10)$$

$$\cdot \sum_{1,1'=0}^{\infty}(-1)^{1'}\sqrt{(21+1)(21'+1)}\begin{pmatrix}1 & k & 1'\\0 & 0 & 0\end{pmatrix}((-1)^{\pi}+(-1)^{1+1'})\,\sigma_0^{k*}(1,1').$$

The 3j-symbols impose restrictions on nonvanishing multipoles: The 3j-symbol inside the 1 sum implies that multipoles with k even (odd) originate exclusively from coherences with 1+1' even (odd) to leading order in 1/n. According to the phase factors in (3.10) π is even (odd) when 1+1' is even (odd). We find therefore the parity of the operator to be related to its rank in such a way that π is even (odd) when k is even (odd). Furthermore, the 3j-symbol in front of (3.10) vanishes if k_1+k_2+k is odd. Recalling that the T-quantum number of the multipoles is given by $(-1)^{k_1+k_2+\pi}$ it follows that only expectation values of T_e operators are non-zero in the limit $n\to\infty$. We finally observe that expectation values (3.10) with different (k_1,k_2) but with the same k become linearly dependent because $k_{1,2}$ do not appear inside the l-sum. A single element is therefore sufficient to construct an operator set in the Rydberg limit. Choosing for simplicity $k_1=k$ and $k_2=o$ and using the reality of (3.10) we arrive at

$$<U_0^k(\pi,k,o)> = \frac{(-1)^k}{n} \sum_{1,1'=o}^{\infty} (-1)^1 \sqrt{(21+1)(21'+1)} \begin{pmatrix} 1 & k & 1' \\ o & o & o \end{pmatrix} \sigma_0^k(1,1') \tag{3.11}$$

with the constraint $k + \pi$ even. This linear dependence can be easily understood by observing that the two pseudospins (3.2) become in the limit $n\to\infty$

$$\vec{j}(1,2) \cong \mp \frac{1}{2} \vec{a} = \mp \frac{n}{2Z_p} \vec{A} \quad . \tag{3.12}$$

The dynamical multipoles may therefore be considered (up to normalization factors) as multipoles of $\vec{a}$ or $\vec{A}$. The latter facilitates the physical interpretation of coherence parameters in the Rydberg limit.

Using now the continuation across the ionization limit the dynamical multipoles (3.11) are expected to describe coherences in the low-lying continuum as well. A partial-wave expansion of the DDCS first discussed by Briggs and Day[15] reads in terms of the density matrix for the continuum $<vlm|\sigma|vl'm'>$ as

$$\frac{d\sigma}{d\vec{v}} = \sum_{1,m,1'm'} (-1)^{1+1'} Y_1^m(\hat{v}) Y_{1'}^{m'}(\hat{v})^* <vlm|\sigma|vl'm'> \delta_{m,m'} \tag{3.13}$$

where $|vlm>$ denotes the partial waves as defined in (2.2) and the phase factor reflects the incoming boundary conditions. Inserting the expansion for a product of spherical harmonics[28] (3.13) becomes

$$\frac{d\sigma}{d\vec{v}} = \frac{1}{4\pi} \sum_{k=o}^{\infty} (2k+1) P_k(\cos\theta) \sum_{1,1',m} ((21+1)(21'+1))^{1/2} \tag{3.14}$$

$$(-1)^{1+1'-m} \begin{pmatrix} 1 & k & 1' \\ -m & o & m \end{pmatrix} \begin{pmatrix} 1 & k & 1' \\ o & o & o \end{pmatrix} <vlm|\sigma|vl'm>$$

with $\theta = \cos^{-1}(\hat{v}\cdot\hat{z})$ the emission angle in the restframe of the projectile. Using the spherical multipoles (3.8) of the density matrix, (3.14) can be rewritten as

$$\frac{d\sigma}{d\vec{v}} = \frac{1}{4\pi} \sum_{k=o}^{\infty} \sqrt{2k+1} P_k(\cos\theta)(-1)^k \sum_{1,1'} (-1)^1 ((21+1)(21'+1))^{1/2} \begin{pmatrix} 1 & k & 1' \\ o & o & o \end{pmatrix} \tag{3.15}$$

$$\cdot \sigma_0^k(1,1') \quad .$$

Comparison with (3.11) shows now that in the zero-velocity limit the right-hand side of (3.15) can be replaced by the Rydberg limit of expectation values for dynamical multipoles $<U_0^k>$. Using the continuity across the ionization limit of the cross section per unit energy interval we find

$$\lim_{v \to o} D(\varepsilon > o)\ \frac{d\sigma}{d\vec{v}} = \lim_{n \to \infty} \frac{D(\varepsilon < o)\ n\ U_0^0(o,o,o)}{4\pi} \sum_{k=o}^{\infty} \sqrt{2k+1} \qquad (3.16)$$

$$\cdot\ P_k(\cos\theta)\ \frac{<U_0^k(\pi,k,o)>}{<U_0^0(o,o,o)>} \quad .$$

If one parametrizes the zero-velocity limit of the DDCS as

$$\frac{d\sigma}{d\vec{v}} = \frac{\sigma}{v} \sum_{k=o}^{\infty} P_k(\cos\theta)\ \beta_k \qquad (3.17)$$

in terms of the isotropic part σ/v of the cross section and of the anisotropy parameters β_k we can identify

$$\sigma = \frac{1}{4\pi}\ \lim_{n \to \infty} D(\varepsilon < o)\ n\ U_0^0(o,o,o) \qquad (3.18a)$$

$$= \frac{1}{4\pi}\ \lim_{n \to \infty} D(\varepsilon < o) \sum_{l,m} <nlm|\sigma|nlm>$$

and

$$\beta_k = (2k+1)^{1/2} \cdot \frac{<U_0^k(\pi,k,o)>}{<U_0^0(o,o,o)>} \quad . \qquad (3.18b)$$

Eq. (3.18) is the desired parametrization of the cusp cross section in terms of the density matrix for the low-lying continuum. Each anisotropy parameter for cusp electrons is associated with certain coherence parameters in the density matrix. For example, β_1 usually referred to as the asymmetry parameter of the cusp, is exclusively determined by partial-wave coherences of mixed parity with $\Delta l = |1-1'| = 1$. By the proportionality between $<U_0^k>$ and $<A_0^k>$ in the Rydberg limit, β_1 corresponds (up to trivial factors) to the expectation value $<-A_z>$. For bound states, on the other hand, $<A_z>$ describes the dipole moment or charge asymmetry of the orbit[20,29,30]. The present analysis exhibits the direct correspondence between the forward-backward asymmetry of the charge distribution for bound states and the forward-backward asymmetry in the cusp electron distribution. This relationship suggests that a cusp shape analysis may reveal informations on the shape of high-lying Rydberg orbits and vice versa.

4. Application to electron capture to continuum (ECC)

In the present section we discuss recent experimental and theoretical results for the DDCS following electron capture to continuum in the light of the foregoing analysis.

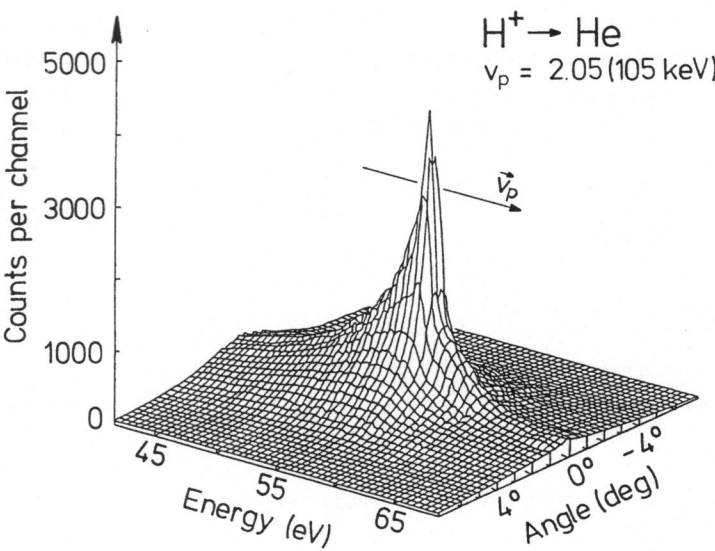

Fig. 3: Experimental electron distribution[5] following the ECC process
H^+ + He → H^+ + He^+ + e.

The most remarkable feature in the DDCS for ECC is the skewness (Fig. 3) towards
lower lab velocities[4,5,21] described by a negative β_1 (and possibly higher β_k with
k odd). Shakeshaft and Spruch[12] have pointed out that the cusp asymmetry is a clear
signature of the presence of second Born terms in the electron capture amplitude since
the first Born approximation to charge transfer (usually referred to as the Oppenhei-
mer-Brinkman-Kramers (OBK) approximation) gives an isotropic electron distribution in
the limit v → o (β_k = o for all k ǂ o). In Fig. 4 the results of Shakeshaft and
Spruch for the singly differential cross section (SDCS)

$$\frac{d\sigma}{dv_e} = v_e^2 \int_o^{\theta_0} \int_o^{2\pi} \sin\theta_e \, d\theta_e \, d\phi_e \left(\frac{d\sigma}{d\vec{v}}\right) \tag{4.1}$$

where θ_0 denotes the half-angle of a cone centered about the forward direction in
laboratory frame are displayed.

The origin of the failure of the OBK approximation to account for the cusp asymmetry can be easily understood. As is evident from a partial-wave decomposition of σ
shown in Fig. 4b the isotropic cusp is <u>not</u> due to a dominance of the s-wave as usually
assumed, except at really asymptotically high projectile velocities. The p- and d-
wave contributions dominate in the present case the zero-velocity spectrum. Instead,
the isotropic cusp results from a lack of partial-wave coherence between waves

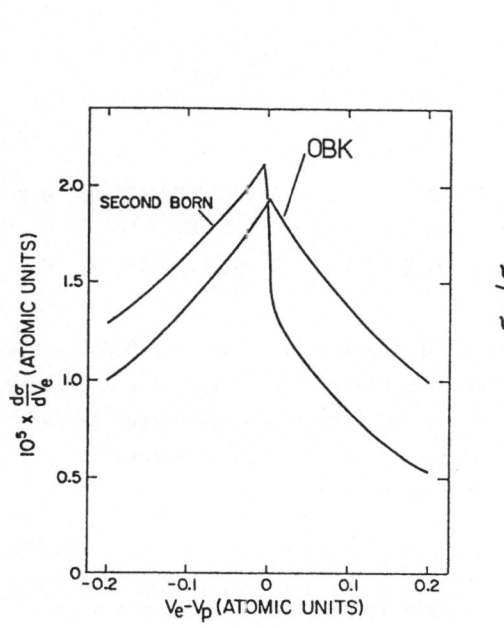

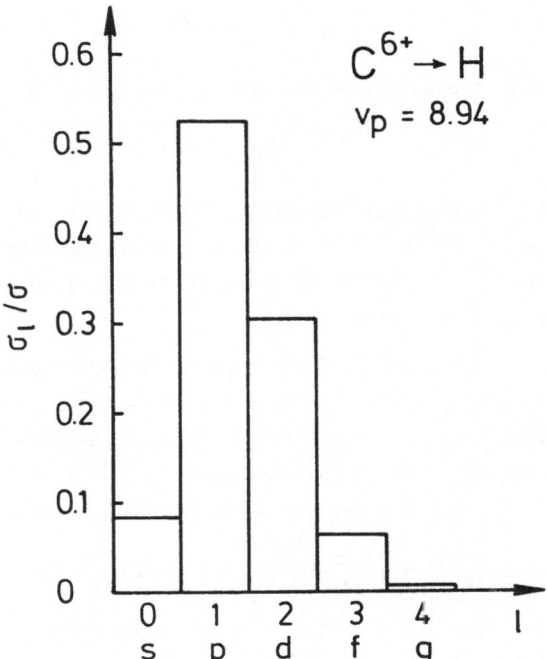

Fig. 4: a) ECC cusp in the SDCS following $C^{6+} \to H$ ($v_p = 8.94$; $\cong$ 2 MeV/nucleon) in OBK and second Born approximation[12].

b) Partial-wave decomposition of the cusp cross section σ in OBK approximation.

of opposite parity. According to (3.11) and (3.18) the asymmetry parameter $\beta_1 \propto \langle U_0^1 \rangle$ belongs to the coherence parameters of class P_oT_e. It is now straightforward to show that in the Born approximation for excitation as well as charge transfer coherences of classes P_oT_e and P_eT_o are absent. For excitation this simply follows from the observation that the transition amplitude can be written as a matrix element of the boost operator[15,16],

$$t_{i \to f}(\vec{K}) = f(K^2) \langle f | \exp(i \vec{K} \vec{r}) | i \rangle \tag{4.2}$$

with $f(K^2)$ independent of the initial and final states. Expanding the exponential as $1 + i \vec{K} \vec{r} + \dots$ it is readily seen that the effective transition operator in Born approximation contains only terms of the class P_eT_e (even powers in $(i \vec{K} \vec{r})$) and of the class P_oT_o (odd powers in $(i \vec{K} \vec{r})$) and thus cannot produce coherences of the classes P_oT_e or P_eT_o. A similar analysis applies to charge transfer. It may also be generalized to the double-series expansion devised by Meckbach, Garibotti, and coworkers [14,31] to account for finite-velocity corrections in the DDCS

$$\frac{d\sigma}{d\vec{v}} = \frac{1}{v} \left(\sum_{k,j=0}^{\infty} B_k^{(j)} v^j P_k(\cos\theta) \right). \tag{4.3}$$

The PT symmetry of the Born approximation gives then the selection rule[27]

$$B_k^{(j)} = o \qquad \text{for k+j odd.} \tag{4.4}$$

Notice that odd Legendre polynomials may contribute for higher-order corrections in v (e.g., $B_1^{(1)}$). This is the origin of the slight asymmetry observed by Chan and Eichler[10] in their Born calculation taking into account first-order corrections in velocity of the final state.

The absence of all even-parity multipoles β_k in the v→o limit of the DDCS for ECC in OBK approximation is not a consequence of the PT symmetry but is specific to electron capture. The final-state wave function enters the expression for charge transfer cross section in OBK approximation[10] by the square of its Fourier transform[23] $|\psi_{\vec{v}}(\vec{q})|^2$. The zero-velocity limit is readily derived to give

$$\psi_{\vec{v}}^-(\vec{q}) \underset{v\to 0}{=} \left(\frac{2\ Z_p^3}{\pi^3 v}\right)^{1/2} \frac{1}{q^4} \exp\left[+ 2i\ \frac{Z_p}{q}\ \hat{q} \cdot \vec{v}\right]. \tag{4.5}$$

The dependence on the angle $\hat{\theta} = \cos^{-1}(\hat{q} \cdot \hat{v})$ is completely contained in a "plane wave" phase factor that drops when inserting (4.5) into the expression for the cross section. The DDCS is therefore isotropic (β_k=o for k≠o) although all partial waves are present in the zero-velocity wave function in momentum space as can be seen from a partial-wave expansion of the plane-wave factor in (4.5).

The symmetry analysis can also be applied[27] to higher-order perturbation theories[32] to charge transfer yielding similar selection rules for certain approximations such as the eikonal approximation and the Continuum Intermediate States(CIS) approximation.

Barrachina and Garibotti[14] have calculated the first four coefficients $B_k^{(j)}$ within a multiple-scattering approach. They find reasonable agreement with experimental data on the cusp shape in H^+ + He collisions. For highly charged argon ions incident on He the semiclassical impulse approximation[13] could reproduce experimental data[4] at v=18 a.u.

5. Electron loss to continuum (ELC)

Electron loss to continuum results from a soft collision in which a projectile electron is excited to a state just above the ionization limit due to an interaction with either the target nucleus or a target electron. In the latter case simultaneous excitation or ionization of the target electron (a doubly inelastic process) takes place. By contrast to electron capture the Born approximation for excitation is generally believed[32] to be valid in the limit of large projectile velocities. As the selection rules (4.4) apply to ELC as well they should be visible in the experimental electron distribution. In turn, occurrence of forbidden multipoles provides clear evidence of higher-order Born terms to be present in the ionization amplitude at finite projectile velocities.

Restricting to the zero-velocity limit, Briggs and Day[15] have found an anisotropic DDCS for ELC for an electron initially in a hydrogenic 1s state in the Born approximation,

$$\frac{d\sigma}{d\vec{v}} = \frac{\sigma}{v} (1+\beta_2 P_2(\cos\theta)) \tag{5.1}$$

containing a second-order multipole parameter $\beta_2 \propto \langle A_0^2 \rangle$. Recent experimental data of Meckbach et al.[17] (Fig. 5) confirm the absence of a significant asymmetry ($\beta_k = 0$ for k odd) in agreement with the prediction of the Born approximation. They exhibit also a preferred longitudinal emission parallel to the beam axis ($\hat{z}$) described by $\beta_2 > 0$ (see Fig. 7). The presence of higher-order even multipoles, in particular β_4, found in the contour lines[17] indicates the presence of higher-order Born terms and awaits further theoretical explanation.

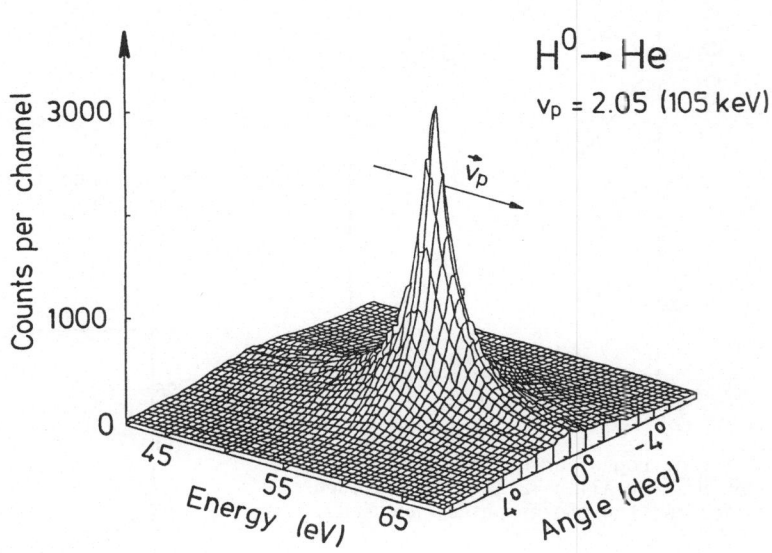

$H^0 \to He$

$v_p = 2.05$ (105 keV)

Fig. 5: Experimental electron distribution[17] following the ELC process H + He → H⁺ + He(?) + e

The Born approximation to ELC has recently been generalized to arbitrary hydrogenic initial states (n'l'm')[16]. The resulting DDCS

$$\frac{d\sigma}{d\vec{v}} = \frac{\sigma (n'l'm')}{v} \sum_{\substack{k=0 \\ (\text{even})}}^{2n'} \beta_k(n'l'm') P_k(\cos\theta) \tag{5.2}$$

becomes highly anisotropic with even multipoles contributing up to order k = 2n'. The DDCS parameters σ and β_k for the ELC process He⁺(n'l'm') + H → He⁺⁺ + H(?) + e and all n'=1 and 2 initial states are shown in Figs. 6-8. Contributions originating from doubly inelastic processes are taken into account within the framework of the closure approximation[33].

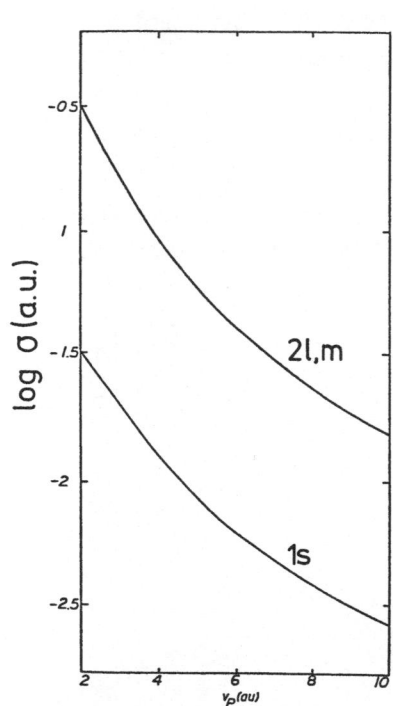

Fig. 6: Log σ for ELC of He⁺(n'l'm') on H. For all n'=2 states σ is identical within the graphical accuracy.

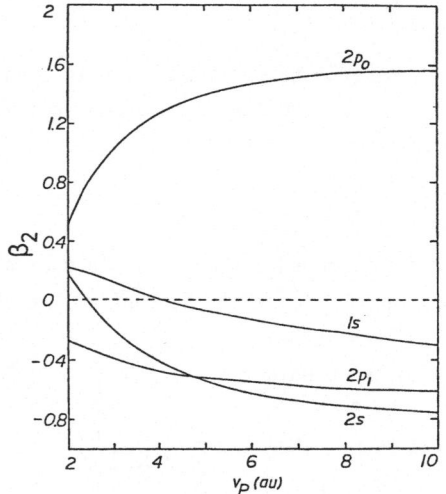

Fig. 7: Second-order anisotropy coefficient β_2(n'l'm') for ELC of He⁺(n'l'm') on H.

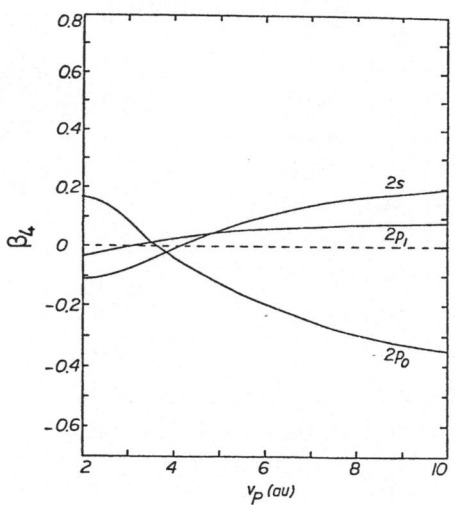

Fig. 8: Fourth-order anisotropy coefficient β_4(n'l'm') for ELC of He⁺(n'l'm') on H

The isotropic part σ is almost independent of the angular momentum quantum numbers of the initial state but shows a sharp rise with increasing principal quantum number, or equivalently, with decreasing binding energy. The latter points to the dominant contribution of distant (soft) collisions to the ejection of low-velocity electrons.

The anisotropy parameters are strongly dependent on the angular momentum quantum numbers of the initial state giving rise to a variety of cusp shapes depending on (l',m') and the projectile velocity. The DDCS for s-states changes its emission pattern from a preferentially longitudinal ejection at lower velocities ($v_p \leq 4$ a.u., see Fig. 5) to a preferred transverse emission ($v_p > 4$ a.u.) at higher velocities (Fig. 9). In

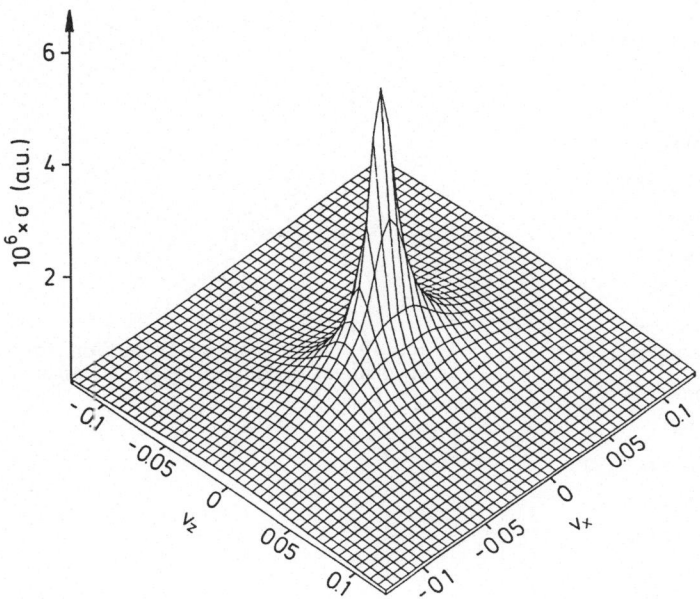

Fig. 9: Electron distribution following the ELC process $H(2s) + He \rightarrow H^+ + He(?) + e$ at $v_p = 10$ a.u. A resolution volume of size ($\Delta v_\parallel = 9.10^{-3}$, $\Delta v_\perp = 1.8 \cdot 10^{-2}$) has been folded into the DDCS.

general, isotropic initial states have negative β_2 values in the limit $v_p \rightarrow \infty$. This is due to the fact that the momentum transfer during the collision is the "source" of anisotropy for isotropic states, and that this momentum transfer is oriented purely transverse with respect to the beam velocity in the limit $v_p \rightarrow \infty$.

For non-isotropic initial states the resulting electron distribution is a complicated mixture of the initial-state anisotropy and the collisionally induced anisotropy. Figs. 10 and 11 display the electron distribution ejected from a $H(2p_o)$ state in collisions with Helium at $v_p = 10$ a.u. The DDCS exhibits a highly singular behaviour with a cusp singularity at the origin (v=o) superposed by a deep "valley" along the line ($v_z = o$, v_x). The latter is obviously a "memory" effect from the nodal plane of the initial state. Depending now on the resolution volume folded into the electron distribution the singular behaviour near the origin allows for drastic variations of the resulting shape. Choosing a cubic volume of linear size $\Delta v = 9.10^{-3}$ a.u. we find a narrow ridge replacing the usual cusp peak (Fig. 10). If we extend the transverse extension of the resolution volume to $\Delta v_\perp = 1.8 \cdot 10^{-2}$ a.u. corresponding to an angular resolution of about $\Delta\theta_e = 0.1°$ in the lab frame (Fig. 11), the central peak has completely disappeared and has been replaced by a valley. It can be shown that all initial states of odd parity with respect to reflections $(z \rightarrow -z)$, i.e. states with (l-m) odd, give inverted cusps in the limit of large projectile velocities. As Böckl et al.[34] have pointed out a similar inversion may appear in the binary-encounter peak in the DDCS.

Eq. (5.2) has recently been applied to electron emission from highly ionized atoms. ELC experiments with C^{q+}, O^{q+} and Si^{q+} ions in the MeV range[4] having a partially filled L shell

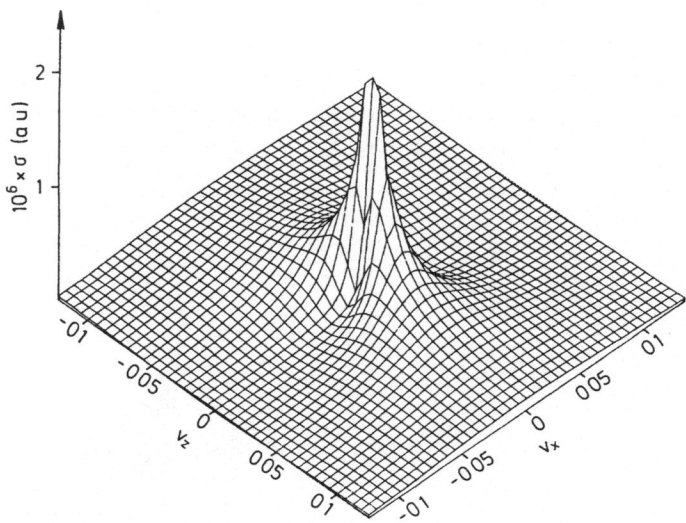

Fig. 10: Electron distribution following the ELC process H($2p_0$) → He at v_p=10 a.u.; $\Delta v_\parallel = \Delta v_\perp = 9 \cdot 10^{-3}$.

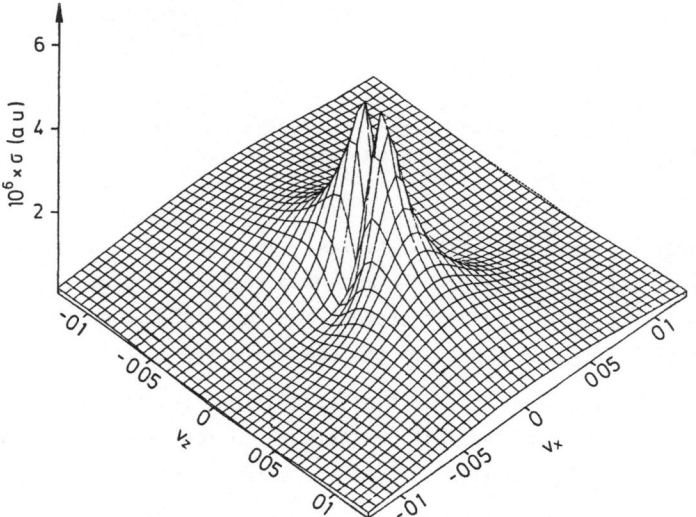

Fig. 11: Electron distribution following the ELC process H($2p_0$) → He at v_p=10 a.u.; $\Delta v_\parallel = 9 \cdot 10^{-3}$ and $\Delta v_\perp = 1.8 \cdot 10^{-2}$.

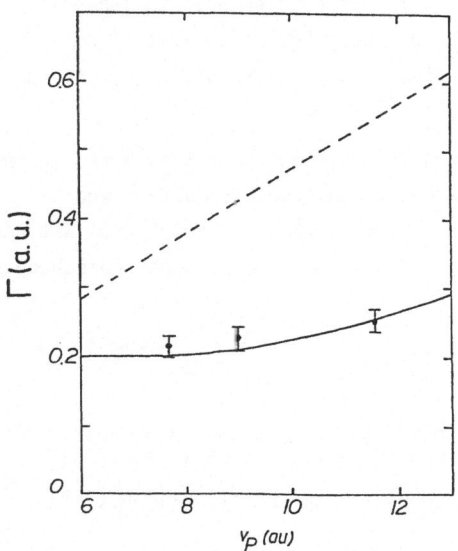

show a narrow and almost v_p independent width Γ (FWHM) of the cusp peak in the singly-differential cross section. Fig. 12 displays the experimental and theoretical results[16] on $\Gamma(v_p)$ for $O^{5+} \to$ Ar. The Li-like electronic configuration is approximated by hydrogenic wave-functions with effective charges. The dominant contribution to the DDCS is due to the ionization of the 2s electron. The highly anisotropic ELC distribution has turned out to be of crucial importance in explaining the narrow cusp width. The strongly preferred transverse emission of the 2s-electron (Fig. 9) narrows the ELC cusp in the SDCS which (approximately) measures the longitudinal velocity distribution. The present theoretical result including β_2 and β_4 contributions is in much better agreement with experimental data for the width than a model for an isotropic cusp ("Dettmann shape"[11]).

Fig. 12: ELC cusp width for O^{5+} on argon ($\theta_0 = 3.14 \times 10^{-2}$ rad $\cong 1.8°$).
—— calculation including β_2 and β_4;
---- width for isotropic electron emission; ($\beta_2 = \beta_4 = 0$)
Φ experimental data (Ref. 4).

6. Conclusions

A group-theoretical classification of various cusp anisotropies has been discussed. The anisotropy parameters β_k in the doubly-differential cross section have been expressed as expectation values of multipoles of the Runge-Lenz operator. The present method taking explicitly into account the dynamical $O(4)$ symmetry of the Coulomb problem applies to hydrogenic projectiles with a Coulombic final-state interaction between the projectile ion and the released electron.

The parametrization of anisotropy parameters in terms of expectation values of $O(4)$ group generators is closely related to a recent classification of bound-states coherences[20]. An interesting feature of the present approach is that the same set anisotropy parameters may describe the population of both high-lying Rydberg states and low-lying continuum states. Relying on the continuity across the ionization limit β_k should become smooth functions in the transition region between bound states and continuum states. This suggests that a cusp shape analysis may reveal informations on the shape of Rydberg orbits. Conversely, recent data on the anisotropic Rydberg population[35] may be directly compared with cusp measurements for the same process when expressed in terms of multipole expectation values.

As this method only relies on the properties of the hydrogenic projectile-electron system but not on specific features of the target it may be applied also to ion-solid collisions. A more complete picture of the physics of near threshold excitation in ion-atom (solid) collisions is expected to emerge from future investigations along these lines.

I should like to thank several colleagues who have contributed to this work. Among those are M.Breinig, S.Elston, and I.Sellin (Oak Ridge), H.Gabriel, E.Kupfer, and H. Schröder (Berlin), and L.Dubê (Freiburg). This work was supported in part by the National-al Science Foundation and by the Sonderforschungsbereich 161 of the Deutsche Forschungs-gemeinschaft.

References
1. C.B.Crooks, M.E.Rudd, Phys.Rev.Lett. 25, 1599 (1970)
2. K.G.Harrison, M.W.Lucas, Phys.Lett. 33A, 142 (1970)
3. V.H.Ponce, W.Meckbach, Comments At.Mol.Phys. 10, 231 (1981)
4. M.Breinig, S.Elston, S.Huldt, L.Liljeby, C.Vane, S.Berry, G.Glass, M.Schauer, I.Sellin, G.Alton, S.Datz, S.Overbury, R.Laubert, M.Suter, Phys.Rev. A25, 3015 (1982)
5. K.O.Groeneveld, W.Meckbach, I.A.Sellin, J.Burgdörfer, Comments At.Mol.Phys. 14, 187 (1984)
6. H.-D.Betz, D.Röschenthaler, J.Rothermel, Phys.Rev.Lett. 50, 34 (1983)
7. Y.Yamazaki, N.Oda, Phys.Rev.Lett. 52, 29 (1984)
8. J.Macek, Phys.Rev. A1, 235 (1970)
9. M.E.Rudd, J.Macek, Case Stud.At.Mol.Phys. 3, 47 (1972)
10. F.Chan, J.Eichler, Phys.Rev. A20, 367 (1979)
11. K.Dettmann, K.G.Harrison, M.W.Lucas, J.Phys. B7, 269 (1974)
12. R.Shakeshaft, L.Spruch, Phys.Rev.Lett. 41, 1037 (1978)
13. D.Jakubassa-Amundsen, J.Phys. B16, 1767 (1983)
14. C.R.Garibotti, J.Miraglia, J.Phys. B14, 863 (1981); R.Barrachina, C.R.Garibotti, Phys.Rev. A28, 1821 (1983)
15. F.Drepper, J.Briggs, J.Phys. B9, 2063 (1976); J.Briggs, F.Drepper, J.Phys. B11, 4033 (1978); M.Day, J.Phys. B13, L65 (1980); J.Briggs, M.Day, J.Phys. B13, 4797 (1980); M.Day, J.Phys. B14, 231 (1981)
16. J.Burgdörfer, Phys.Rev.Lett. 51, 374 (1983); J.Burgdörfer, M.Breinig, S.Elston, I.Sellin, Phys.Rev. A28, 3277 (1983)
17. W.Meckbach, R.Vidal, P.Focke, I.B.Nemirovsky, E.Gonzales Lepera, Phys.Rev.Lett. 52, 621 (1984)
18. S.Elston et al., this conference
19. E.P.Wigner, Phys.Rev. 73, 1002 (1948)
20. J.Burgdörfer, Z.Phys. A309, 285 (1983)
21. M.Rødbrø, F.Andersen, J.Phys. B12, 2883 (1979)
22. M.Lucas, W.Steckelmacher, J.Macek, J.Potter, J.Phys. B13, 4833 (1980)
23. H.Bethe, E.Salpeter, Quantum Mechanics of One- and Two Electron Atoms (Springer-(Verlag, Berlin, 1957)
24. M.Englefield, Group Theory and the Coulomb Problem (Wiley-Interscience, New York, 1972)
25. J.Bosse, H.Gabriel, Z.Phys. 266, 283 (1974) and references therein
26. U.Fano, J.Macek, Rev.Mod.Phys. 43, 553 (1973)
27. J.Burgdörfer, to be published
28. D.Brink, G.Satchler, Angular Momentum (Clarendon Press, Oxford, 1968)
29. C.Havener, W.Westerveld, J.Risley, N.Tolk, J.Tully, Phys.Rev.Lett. 48, 926 (1982)
30. J.Burgdörfer, L.Dubê, Phys.Rev.Lett. (1984), in press
31. W.Meckbach, I.B.Nemirovsky, C.Garibotti, Phys.Rev. A24, 1793 (1981)
32. J.Briggs, J.Macek, K.Taulbjerg, Comments At.Mol.Phys. 12, 1 (1982)

33. J.Lodge, J.Phys. B2, 322 (1969)
34. H.Böckl, R.Spies, F.Bell, D.Jakubassa-Amundsen, Phys.Rev. A29, 983 (1984)
35. E.Kanter, D.Schneider, Z.Vager, D.Gemmell, B.Zabransky, G.Yuanzhuang, P.Arcuni, P.Koch, D.Mariani, W.Van de Water, Phys.Rev. A29, 583 (1984)

A TIME DEPENDENT SECONDARY ELECTRON TRANSPORT MODEL[**]

J. Devooght, J.C. Dehaes, A. Dubus[*] and N. Hollasky
Service de Métrologie Nucléaire , C.P. 165,
Université Libre de Bruxelles
50, avenue F.D. Roosevelt, B-1050 Bruxelles, Belgium

1. Introduction

The purpose of this paper is to give an analytical treatment of secondary emission which is both realistic and reasonably free of some of the shortcomings of the previous models. The main improvement is the time dependence which is absent in all previous models and which is justified here by its potential use in beam-foil experiments. The ion leaving the foil is accompanied by convoy electrons and by secondary electrons. These electrons influence the light emitted by the ion through a transient Stark effect. The difficult problem of evaluating the electric field will not be examined here but the first step of this problem is the knowledge of the distribution of outgoing secondary electrons. A first test of the adequacy of the model is its ability to predict the main characteristics of secondary electron emission.

2. The model

Analytical solutions of electron (or neutron) transport are readily available in the constant speed approximation. However the gain resulting from a correct treatment of boundary conditions and transport effects near the surface are more than offset by the inadequate energy treatment. Moreover the importance of low energy elastic scattering [1,2] which was not recognized in early models makes the P1 or diffusion theory a reasonable approximation [9,10]. However the boundary conditions will be satisfied only approximately although in a better way than in earlier models which all, more or less explicitly, assumed that the energy and angle spectrum is unaffected by the presence of the vacuum.

---

[*] Aspirant F.N.R.S.

[**] Work supported by the I.I.S.N.

Since we forego the assumption of constant mean free path, the only analytical method is through first order differential equations, i.e. the age theory or genera-
lizations thereof through synthetic scattering kernels. It is true that considerable progress has been obtained (see Williams[3]) in transport problems with power-
law mean free paths by means of Mellin transforms but such simple laws, although
adequate for sputtering or ion slowing down are inadequate for electron transport
in the 0-50 ev range where the mean free path has a pronounced minimum[4]. We
chose therefore to use synthetic kernels that allow arbitrary total cross-sections,
arbitrary mean energy losses and arbitrary mean angular deflections and even higher
moments.

We start from the Boltzmann equation

$$(\frac{1}{v} \frac{\partial}{\partial t} + \bar{\Omega}\bar{\nabla})\phi(\bar{r},E,\bar{\Omega},t) + \Sigma_s(E)\phi(\bar{r},E,\bar{\Omega},t) = \int [\Sigma_s(E'\to E,\bar{\Omega}'\to\bar{\Omega}) + \Sigma_s^S(E'\to E,\bar{\Omega}'\to\bar{\Omega})]$$

$$\phi(\bar{r},E',\bar{\Omega}',t)dE'd\bar{\Omega}' + S(\bar{r},E,\bar{\Omega},t) \tag{1}$$

where $\Sigma_s^S(E'\to E,\bar{\Omega}'\to\bar{\Omega})$ is the production cross-section for electrons out of the Fermi
sea, decomposed into

$$\Sigma_s^S(E'\to E,\bar{\Omega}'\to\bar{\Omega}) = \Sigma_{e,s}^S(E'\to E,\bar{\Omega}'\to\bar{\Omega}) + \Sigma_{p,s}^S(E'\to E,\bar{\Omega}'\to\bar{\Omega}) \tag{2}$$

where $\Sigma_{e,s}^S$ and $\Sigma_{p,s}^S$ are respectively the production cross-section by inelastic electron-electron scattering and by plasmon decay.
Moreover

$$\Sigma_s(E') = \iint \Sigma_s(E'\to E,\bar{\Omega}'\to\bar{\Omega})dEd\bar{\Omega} \tag{3}$$

Similarly $\qquad\qquad S = S_e + S_p \tag{4}$

with the same meaning for the symbols.
Cross-sections Σ_s and Σ_s^S have been computed for electron gas by Ritchie and his
collaborators[4,5],Chung and Everhart[6], Rösler and Brauer [7].
Exact solutions of (1) being analytically impossible for realistic cross-sections,
we shall use the technique of synthetic kernels widely used in neutron transport
theory : exact cross-sections are converted into approximate cross-sections which
allow the conversion of the equation (1) into a solvable partial differential
equation. After expansion of the electron flux in Legendre polynomials

$$\phi(x,E,\mu,t) = \overset{N}{\underset{o}{S}} \frac{2\ell+1}{2} P_\ell(\mu)\phi_\ell(x,E,t) \tag{5}$$

we obtain the "P_N system" of N+1 integrodifferential equations [14].
From rotational invariance

$$\Sigma_s(E' \to E, \bar{\Omega}' \to \bar{\Omega}) = \Sigma_s(E' \to E, \bar{\Omega}' \cdot \bar{\Omega}), \text{ where } \bar{\Omega}' \cdot \bar{\Omega} = \mu_o$$

In plane geometry $\Sigma_s(E' \to E, \mu' \to \mu) = \int_o^{2\pi} \Sigma_s(E' \to E, \mu_o) d\phi$

The cross-section can be developed in the standard way :

$$\Sigma_s(E' \to E, \mu_o) = \sum_o^{\infty} \frac{2\ell+1}{4\pi} B_\ell(E' \to E) P_\ell(\mu_o) \tag{6}$$

where for $E' \gg E_F$ the series is only slowly converging, the total cross-section is :

$$\Sigma_s(E) = \int_o^E B_o(E \to E') dE' \tag{7}$$

We adopt the following synthetic kernel :

$$\Sigma_s^*(E' \to E, \mu' \to \mu) = \sum_o^N \frac{2\ell+1}{2} B_\ell^*(E' \to E) P_\ell(\mu') P_\ell(\mu) + C(E' \to E) \delta(\mu - \mu') \tag{8}$$

and a similar development for $\Sigma_s^{*S}(E' \to E, \mu' \to \mu)$

with $B_\ell^*(E' \to E) = k(E', E) A_\ell(E') + K_\ell(E') \delta(E - E')$ $\qquad$ (9)

$$C(E' \to E) = k(E', E) P(E') \triangleq \frac{H(E' - E) H(E - E_F)}{E' - E_F} P(E') \tag{10}$$

and similar developments for $B_\ell^{*S}(E' \to E)$ and $C^S(E' \to E)$. E_F is the Fermi energy. The choice of the denominator of (10) is dictated by analogy with Wolff's treatment to allow for the exclusion of the Fermi sphere [11]. If N=1, we have 10 arbitrary functions reduced to 5 because Σ_s and Σ_s^S appear always as a sum.

The use of P1 approximation is justified by the fact that elastic scattering dominates all other processes and the elastic scattering is reasonably isotropic at low energies (< 100 ev). Moreover the "extended P1 transport approximation" of Wienke [9] and Morel [10] can take into account scattering cross-sections and sources of type P2 within the structure of P1 equations, by means of $C(E' \to E)$ and $C^S(E' \to E)$.

Defining the angular and energy moments :

$$M_{\ell m}(E') \triangleq 2\pi \int_{-1}^{+1} P_\ell(\mu_o) d\mu_o \int_{E_F}^{E'} (E' - E)^m \Sigma_s(E' \to E, \mu_o) dE \tag{11}$$

and similarly for $M_{\ell m}^S(E')$, we equate the exact moments with the moments of the synthetic kernels. We equate the exact moments with the moments of the synthetic kernels (8) for $(\ell, m) = (0,0), (0,1), (1,0), (1,1), (2,0)$ to obtain the 5 functions. The mathematical details of the model are left to a more detailed paper.

We define the average relative energy loss (or lethargy gain) by

$$\frac{2}{E'-E_F} M_{o,1}^{(s)} (E') \overset{\Delta}{=} \xi^{(s)} (E') \Sigma_s^{(s)} (E') \tag{12}$$

and we define

$$\bar{\xi}(E') \Sigma (E') \overset{\Delta}{=} [\xi(E') - \overline{\mu \xi}(E')] \Sigma_s (E') + [\xi^s (E') - \overline{\mu \xi}^s (E')] \Sigma_s^s (E') \tag{13}$$

a total scattering cross-section

$$\Sigma(E') \overset{\Delta}{=} \Sigma_s (E') + \Sigma_s^s (E') \tag{14}$$

a transport cross-section

$$(3D(E'))^{-1} = \Sigma_{tr}(E') \overset{\Delta}{=} \Sigma_s (E') - \mu(E') \Sigma_s (E') - \mu^s (E') \Sigma_s^s (E') + <P_2>(E') \Sigma_s (E')$$

$$+ <P_2>^s (E') \Sigma_s^s (E') \tag{15}$$

The final equation for $\phi_o (x,E,t)$ is

$$-D(E) \frac{\partial^2 \phi_o}{\partial x^2} (x,E,t) + (\frac{1}{v} \frac{\partial}{\partial t} + \Sigma_s (E) + (\bar{\xi}(E)-1)\Sigma(E) + <P_2>\Sigma_s (E) + <P_2>^s \Sigma_s^s (E)) \phi_o (x,E,t)$$

$$= \int_E^{E_o} \frac{\bar{\xi}(E') \Sigma(E')}{E'-E_F} \phi_o (x,E',t) dE' + S_o (x,E,t) - S_2 (x,E,t) - 3D(E) \frac{\partial}{\partial x} [S_-(x,E,t)$$

$$- S_2 (x,E,t)] \tag{16}$$

To derive (16) a term $\partial^2 \phi_o / \partial t^2$ has been neglected : wave propagation effects are negligible compared to diffusion effects when elastic scattering is dominating all other types of scattering. The extension to arbitrary geometry is straightforward from (16).

Contribution of the last term of (16) is negligible for ions.

A Green's function for an infinite medium source $S(\bar{r},E,t) = \delta(\bar{r}-\bar{r}_o) \delta(t) S(E)$ can be easily found from (16) which is a diffusion equation with an equivalent isotropic source $S_o - S_2$ (which is equivalent to a Greuling-Goertzel model [3]).

We have

$$\phi_o (\bar{r},E,t) = vS(E) \frac{e^{-r^2/4vD(E)t}}{(4\pi vD(E)t)^{3/2}} e^{-v\Sigma_o(E)t} + \int_E^{E_o} \frac{\bar{\xi}\Sigma(E')}{\Sigma_o(E')}$$

$$\frac{e^{-v\Sigma_o(E)(t-\hat{t}_o(E' \to E)) + q(E' \to E)}}{E'-E_F} \frac{e^{-r^2/4(vD(E)(t-\hat{t}_o(E' \to E)) + \hat{\tau}(E' \to E))}}{(4\pi(vD(E)(t-\hat{t}_o(E' \to E) + \hat{\tau}(E' \to E))^{3/2}}$$

$$vS(E') H(t-\hat{t}_o(E' \to E)) dE' \tag{17}$$

where

(a) $\bar{\tau}(E' \to E) \triangleq \dfrac{D(E')}{\Sigma_o(E')} + \int_E^{E'} \dfrac{dE''}{E''-E_F} \dfrac{D(E'')}{\Sigma_R(E'')}$ is the electron age

(b) $t_o(E' \to E)$ is the slowing down time

(c) $\Sigma_o(E) = \Sigma_s(E) + (\bar{\xi}(E)-1)\Sigma(E) + \langle P_2 \rangle \Sigma_s(E) + \langle P_2 \rangle^s \Sigma_s^s(E)$

(d) $q(E' \to E) \triangleq \int_E^{E'} \dfrac{\bar{\xi}\Sigma(E'')}{\Sigma_o(E'')} \dfrac{dE''}{E''-E_F}$

(e) $\Sigma_R(E) \triangleq \dfrac{\Sigma_o^2(E)}{\bar{\xi}\Sigma(E)}$

The first term of (17) is the uncollided contribution; as usual an improved result is obtained as a transport kernel is substituted to the diffusion kernel. The second term is the slowing down term. No electron can appear at energy E before the slowing down time $t_o(E' \to E)$.

Our next task is to convert infinite medium solutions into half-space solutions. The approximate boundary conditions are obtained by writing the conservation of the partial currents, taking into account the fact that electrons outside the escape cone are reflected [7] . If we define U_o the potential barrier and

$$\mu_c^2(E) \triangleq \dfrac{U_o}{E} \tag{18}$$

than the flux satisfies a mixed condition at the plane boundary

$$\dfrac{\partial \phi}{\partial z}(o,E,t) + h(E)\phi(o,E,t) = 0 \tag{19}$$

with $$h(E) = \dfrac{1}{2D(E)} \left(\dfrac{1 - \mu_c^2(E)}{1 + \mu_c^3(E)} \right) \tag{20}$$

Using (20) the half-space solution $S(z,z_o,\tau) \dfrac{e^{-\rho^2/4\tau}}{(4\pi\tau)}$ is substituted to each term $\dfrac{e^{-[\rho^2+(z-z_o)^2/4\tau]}}{(4\pi\tau)^{3/2}}$ with $\rho^2 = (x-x_o)^2 + (y-y_o)^2$ and

$$S(z,z_o,\tau) = \dfrac{e^{-(z-z_o)^2/4\tau} + e^{-(z+z_o)^2/4\tau}}{(4\pi\tau)^{1/2}} - h\ \mathrm{erfc}\left(\dfrac{z-z_o}{2\sqrt{\tau}} + h\sqrt{\tau}\right)e^{h(z-z_o)+h^2\tau} \tag{21}$$

The next step in the evaluation of secondary electron emission is the evaluation of the energy deposited by the primary electron or ion. As a first test to the theory we have studied the steady state production of secondary electrons resulting from a continuous stream of primary particles. In order to evaluate the time dependent electron distribution accompanying an outgoing ion, it is sufficient to consider a point source $S(\bar{r},E,t) = S(E)\delta(x)\delta(y)\delta(z-v_i t)$ where $S(E)$ is the energy spectrum of the source, v_i the ion velocity assuming the ion crosses the boundary at $t=0$. Results are obtained through lengthy but straightforward analytical calculations assuming energy is deposited according to a law [8,13]

$$D(z_o,E_{po}) = NS_{in}(E_{po})g\left(\frac{z_o}{r_o}\right) \tag{22}$$

When r_o is the range of the primary particle of energy E_{po} and stopping power $NS_{in}(E_{po})$, the stationary outgoing angular distribution is given by (1 : vacuum, 0 : medium)

$$\phi(E_1,\mu_1|E_{po}) = \frac{\delta(E_1|E_{po})}{2h(E_1+U_o)D(E_1+U_o)} \frac{E_1}{E_1+U_o}\left[1 + \frac{3}{2}\sqrt{\frac{E_1\mu_1^2+U_o}{E_1+U_o}} \frac{1-\mu_c^2(E_1+U_o)}{1+\mu_c^3(E_1+U_o)}\right] \tag{23}$$

where $\delta(E_1|E_{po})$ is the total number of secondary electrons of energy E_1 :

$$\delta(E_1|E_{po}) = S_{j} \int_{\mu_c(E_1+U_o)}^{1} 2\pi\mu d\mu \int_o^1 g(y)\Sigma_j(E_p(yr_o)) \frac{r(E_{po})}{S_{in}(E_p(y_{ro}))} S_{in}(E_{po})dy$$

$$\int_{U_o}^{E_p(z_o)} \phi_j(o,\bar{\Omega},E_1+U_o|z_o,E_{el})dE_{el} \tag{24}$$

The sum in (24) is over all production processes, here electron-electron scattering and plasmon decay; ϕ_j is the angular spectrum at the boundary produced by a point anisotropic source of normalized intensity in z_o.

$$\Sigma_{j,s}^S(E_p\rightarrow E_{el},\bar{\Omega}_p\rightarrow\bar{\Omega}_o) / \int\int \Sigma_{j,s}^S(E_p\rightarrow E_{el},\bar{\Omega}_p\rightarrow\bar{\Omega}_o)dE_{el}d\bar{\Omega}_o$$

which allows to compute S_o-S_2 of eq. (16).

The primary particle is assumed to lose energy as a function of depth z_o according to a law $E_p = E_p(z_o)$. The atomic stopping power by process j is $S_{in,j}(E_p)$ and $S_j S_{in,j}(E_p) = S_{in}(E_p)$.

3. Evaluation of the main parameters

3.1. Cross-sections

The model is dependent upon the choice of particular cross-sections only through parameters involved in (17) : the total cross-sections, the energy and angular moments, the parameters derived thereof being evaluated for each process.

The elastic cross-section is obtained from the muffin-tin model used by Lanteri and al. [12] , i.e. the total cross-section is obtained from ref. [2] and the screening parameter η deduced from Σ_{el} is used to evaluate the angular moments by Spencer's formula [13] . The differential inelastic cross-sections used to evaluate the energy and angle moments are obtained from the RPA model of Ritchie [4,5] . The moments are obtained by numerical quadrature.

3.2. The source

To obtain the source we must in principle solve the primary electron transport.

To avoid to treat explicitly this problem, we have related the source to the energy deposition law, where according to Spencer [13] we have a "universal law" $D(E_{po}, z_o) = S_{in}(E_{po}) g(z_o/r_o)$ when S_{in} is the stopping power and r_o the range. The empirical data are normalized by $g(o) = 1, g(1) = 0$ and $E_{po} = \int_o^{r_o} D(E_{po}, z_o) dz_o$. For ease of calculation, electrons resulting from plasmon decay were assumed to be emitted isotropically and with a parabolic spectrum with limits taken from ref.[7] .

3.3. Some results

Sample results are given below for Al where cross-sections have been checked with those of ref. [4,5] .

As an example of the transport coefficients used in eq. (16) D(E) is shown in fig.(1) where the broad minimum at 50 ev corresponds to the minimum of the total mean free path. $<\mu>$ and $<P_2(\mu)>$ are shown in fig. (2) for the production cross-section : the anisotropy is important at low energy (100 ev) but decreases at higher energy. The opposite behaviour is observed for the inelastic diffusion cross-section which is strongly anisotropic at higher energy.

The depth dependence of the energy distribution of secondary electrons are given in fig. (3). The decrease of the mean energy when the depth increases has been observed by Koshikawa and Shimizu [15] in Monte Carlo computations for copper and corresponds to a cooling of the spectrum due to the increase of collisions. Different cross-sections do not allow yet a more quantitative comparison with their results. Fig. (4) shows that the depth dependence normalized to unity shows very little dependence upon the primary energy in the range 0.2 to 1 kev, with a quasi exponential behaviour (characteristic length ∿ 15 Å compared to 18 Å for copper [15]).

The energy spectrum of the current of outgoing secondary electrons is given in fig. (5) and compared very well with experimental results of Roptin [16] and Everhart [17]. Agreement is better than with Schou's theory [8] owing maybe to the fact that many simplying assumptions about cross-sections and transport have not been made. The angular distribution (23) when integrated over energy is very close to isotropic as universally observed.

4. Conclusions

A model has been developed which gives explicit formulation of the energy, radial, time and angular dependence of outgoing secondary electrons, some of which were previously available only through Monte Carlo simulation. The cross-sections are arbitrary as well as the source.

The model has been tested by evaluating some static characteristics and comparing them with other available experimental and theoretical results. The general agreement is fairly good. Numerical results for time dependence will be investigated next.

References

1. P. Sigmund, S. Tougaard, Springer Ser. Chem. Phys. 17, Springer Verlag (1981).
2. J.P. Ganachaud, M. Cailler, Surf. Sci. 83, 498 (1979).
3. M.M.R. Williams, Progress in Nuclear Energy, 3, 1-65 (1979).
4. C.J. Tung, R.H. Ritchie, Phys. Rev. B.16, 4302 (1977).
5. J.C. Ashley, C.J. Tung, R.H. Ritchie, Surf. Sci., 81, 409 (1979).
6. M.S. Chung, T.E. Everhart, Phys. Rev. B.15, 4699 (1977).
7. M. Rösler, W. Brauer, Phys. Stat. Sol., 104 (I), 161-175 (II), 575-587.
8. J. Schou, Phys. Rev., B.22, 2141 (1980).
9. B.R. Wienke, J. Quant. Spectrosc. Radiat. Transfer., 22, 301 (1979).
10. J.E. Morel, Nucl. Sc. Eng., 71, 64 (1979).
11. P.A. Wolff, Phys. Rev., 95, 56 (1954).
12. H. Lanteri, R. Bindi, P. Rostaing, J. Phys. D. Appl. Phys. 13, 677-92 (1980).
13. L.V. Spencer, Phys. Rev. 98, 1597 (1955).
14. B. Davison, Neutron Transport Theory, Oxford University Press (1957).
15. T. Koshikawa, R. Shimizu, J. Phys. D. Appl. Phys. 7, 1303-1315 (1974).
16. D. Roptin, Thesis (University of Nantes, 1975) (unpublished).
17. T.E. Everhart, N. Saeki, R. Shimizu and T. Koshikawa, J. Appl. Phys. 47, 2941 (1976).

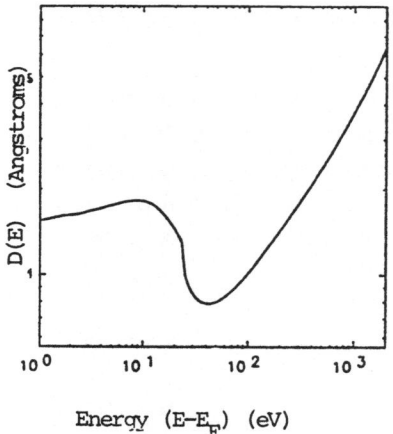

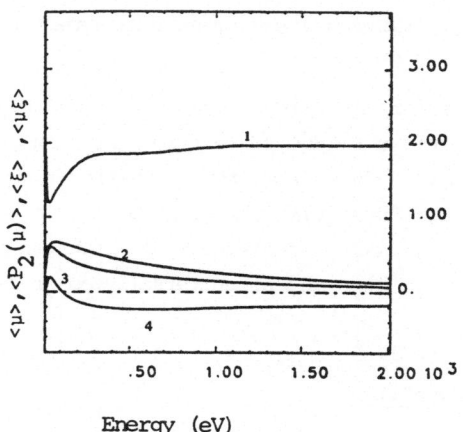

Figure 1 : Diffusion coefficient D(E) : energy dependence.

Figure 2 : It represents some moments for the electron-electron collision creation cross-section.
1 : $\langle\xi\rangle$; 2 : $\langle\mu\xi\rangle$; 3 : $\langle\mu\rangle$;
4 : $\langle P_2(\mu)\rangle$

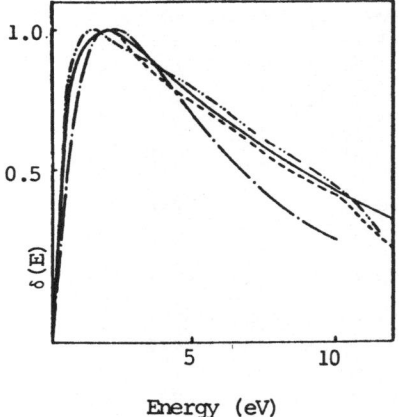

Figure 3 : Energy spectrum $(d\delta/dE)$ of the secondary electrons for perpendicular incidence of 0,8 and 1 keV electrons on aluminium.
Experimental results : (- - -)Ref. 10: Expt. Roptin (0.8 keV);(- •• -) Ref. 11 : Expt. Everhart and al. (1 keV).
Theoretical results : (-• -)Ref. 12 : Schou's theory; (——) this paper.

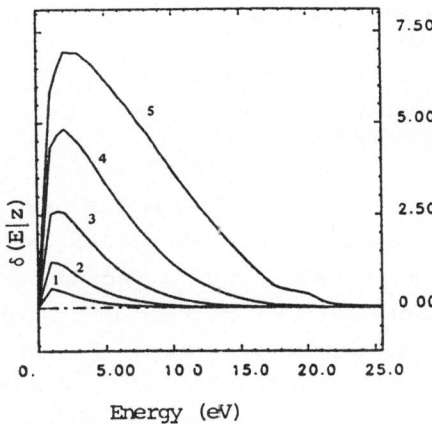

Figure 4 : Depth dependent energy distributions. Depths are 10,20,30, 40 and 50 Å. We can remark the decrease of the maximum energy and of the mean energy with depth increase.

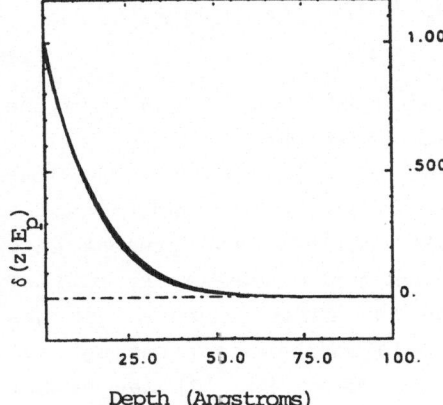

Figure 5 : Primary energy dependent depth distributions. Primary energies are 1000,500 and 200 eV. The differences are very small and the decrease is quasi exponential.

Curves are normalized to the maximum.

CONTINUUM-ELECTRON CAPTURE BY 25-250-keV PROTONS IN HELIUM

P. Dahl
Institute of Physics, University of Aarhus
DK-8000 Aarhus C, Denmark

Electron capture to the continuum in the H^+-He collision has been investigated for impact velocities u = 1-3 a.u., where high yields are obtained. Forward-cone spectra have been measured, and in a second experiment, doubly differential cross sections $\sigma(E,\theta) = d^2\sigma/dEd\Omega$ have been measured at small ejection angles θ different from zero. The aim has been to get information on the cross section $\sigma(E',\theta') = d^2\sigma/dE'd\Omega'$ in the rest frame of the projectile at the limit E' = 0 between capture and ionization. Here, it takes form of a singly differential cross section c_o times an angular distribution,

$$\sigma(E'\theta') \simeq c_o(4\pi)^{-1}W_o(\theta'), \qquad E' \ll 1Ry. \qquad (1)$$

In both experiments, a rather fine collimation is chosen so that the transformation of doubly differential cross sections may be applied to the data fairly close to the singularity in $\sigma(E,\theta)$, corresponding to E' = 0. At finite but not too high electron velocities v' in the projectile frame, $\sigma(E',\theta')$ may be given by

$$\sigma(E',\theta') \simeq c_o(4\pi)^{-1}W_o(\theta') + c_1(4\pi)^{-1}W_1(\theta')\cdot\frac{v'}{u}, \qquad (2)$$

which may be used for an extrapolation to the limit v' = 0 from the data region with differential experimental conditions.

In the forward-cone experiment, one has θ' = 180° (0°) on the low-(high)-energy side of the $\sigma(E)_{\theta_o}$ cusp with differential conditions for $v' \gtrsim 5u\theta_o$, where θ_o is the cone semiangle, which in the present case is 0.38°. In Fig. 1, the projectile-frame cross section is plotted against v'/u for θ' = 180°, 0°. Outside the shaded region, one has v' > $5u\theta_o$. The observed, nearly linear relationship confirms Eq. (2) so that the first term on the right-hand side of Eq. (2) may be obtained for θ' = 180°, 0° by extrapolation of the linear sections to v'/u = 0, and the coefficients to v'/u are obtained as the slopes of the data curves. As an important measure of the anisotropy near v' = 0, the backward-to-forward ratio $W_o(180°)/W_o(0°)$ is derived and plotted against the impact energy in Fig. 2.

In the small-angle experiment with an angular resolution of 0.2°, differential experimental conditions are obtained for $\theta \gtrsim 1°$. If Eq.(1) is fulfilled, it is easily shown that the maximum of $\sigma(E,\theta)$ recorded with a fixed θ is obtained at E = E_p given by

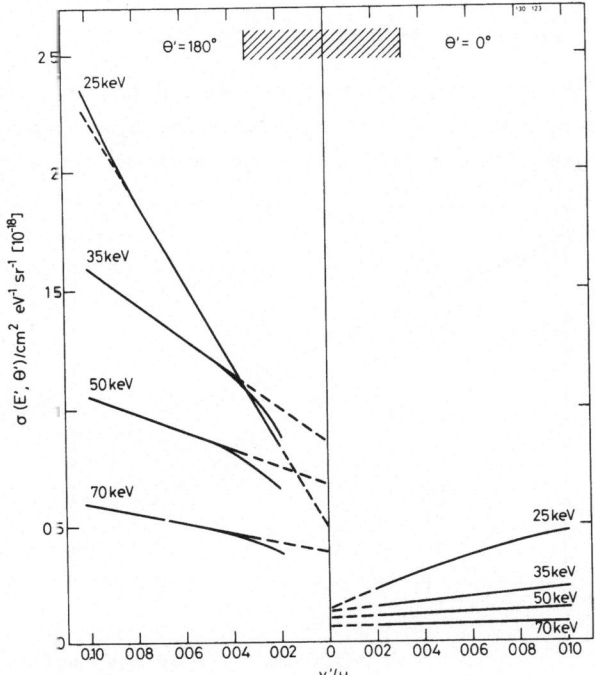

Fig. 1. *Cross sections* σ(E',θ') *with* θ' = 180°,
0° *in the projectile frame obtained from
the forward-cone data. Differential experimental
conditions* v' > 5uθ₀ *outside the shaded region.*

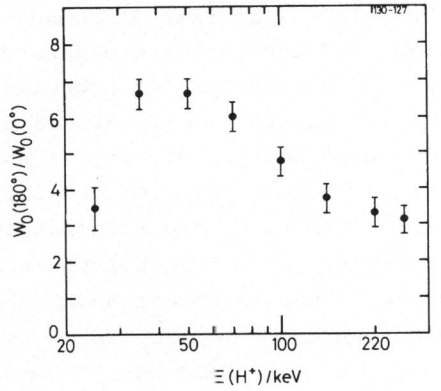

Fig. 2. *Backward-to-forward ratio* W₀(180°)/
W₀(0°) *at the limit* v' = 0.

$$E_p \simeq \epsilon(1 + 2\theta \cot\theta'_p) , \qquad (3)$$

where ϵ is the mass-reduced proton energy, and θ'_p is the value of the projectile-frame angle θ', for which maximum is obtained of $d^2\sigma_0/dE'd\theta'$ given by

$$\frac{d^2 \sigma_o}{dE' d\theta'} = 2\pi \sin\theta' \cdot c_o (4\pi)^{-1} W_o(\theta') \ . \tag{4}$$

Thus, the shift $\varepsilon - E_p$ of the peak position should be proportional to θ. This is confirmed by data shown in Fig.3, where E_p/ε is plotted against θ for impact energies 100 and 200 keV. The derived results for θ'_p are

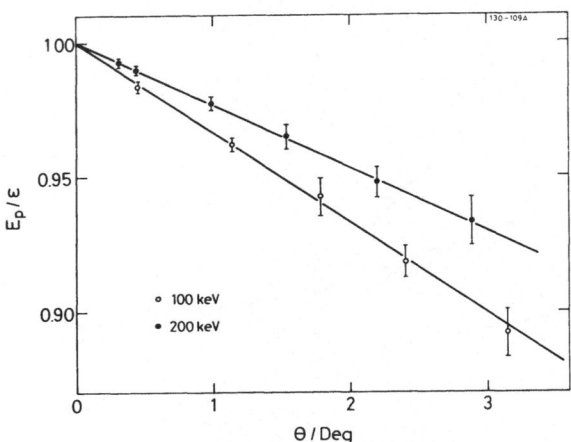

Fig. 3. *Relative peak positions E_p/ε of the small-angle spectra $\sigma(E,\theta)$ with fixed angles θ.*

$$\theta'_p = \begin{cases} 133.6° \pm 1.5° , & 100 \ \text{keV} \\ 124.0° \pm 1.5° , & 200 \ \text{keV} \end{cases} . \tag{5}$$

Furthermore, if Eq. (1) were fulfilled, $c_o(4\pi)^{-1} W_o(\theta')$ would be obtained directly by plotting the data for $\sigma(E',\theta')$ against θ'. However, the data exhibit a rise for θ' approaching 180° and incidate backward-to-forward ratios much higher than the ratios $W_o(180°)/W_o(0°)$ obtained in the forward-cone experiment. This is interpreted as an effect of the dependence of $\sigma(E',\theta')$ on v' since v' increases rapidly for θ' approaching 180° or 0°. In principle, $\sigma(E',\theta')$ data could be obtained in these measurements for fixed values of θ' so that an extrapolation to the limit $v' = 0$ could be performed for each θ', but the present uncertainties prevent this procedure. Instead, the information obtained at $\theta' = 180°$ and 0° in the forward-cone experimentis used. Here, the slopes $d\sigma(E', \theta')/d(v'/u)$ are nearly equal for $\theta' = 180°$ and 0° in a small velocity range $v'/u < 0.1$, but for higher velocities, the slope increases with v'/u for $\theta' = 180°$, while $\sigma(E',\theta')$ becomes nearly independent of v'/u for $\theta' = 0°$. The dependence on v'/u at $\theta' = 180°$ is then used for all angles θ' down to the angle in the forward direction where v'/u exceeds 0.1. This gives an estimate of $c_o(4\pi)^{-1} W_o(\theta')$, with an acceptable backward-to-forward ratio. Inserting this in Eq.(4), the results for $d^2 \sigma_o/dE'\theta'$ presented in Fig. 4 are obtained.

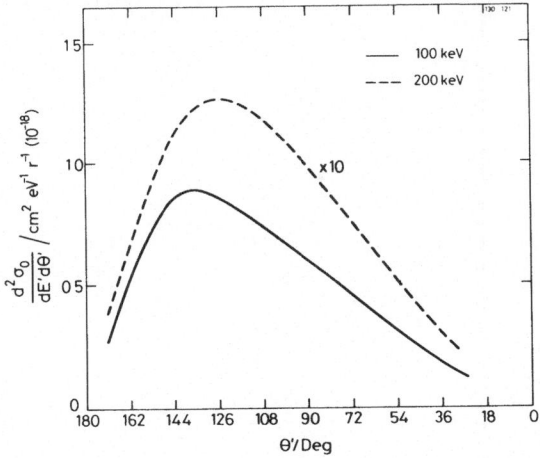

*Fig. 4. Estimate of $d^2 \sigma_o/dE'd \theta'$ at the limit
$v' = 0$ obtained from small-angle data.*

It is noted that the maximum occurs at an angle θ_p', in fair agreement
with Eq. (5), which indicates that the method based on peak positions
may be rather insensitive to the effect of the dependence of $\sigma(E'\theta)$
on v'. The angle θ_p' and the ratio $W_o(180°)/W_o(0°)$ are important figures characterizing the angular distribution $(4\pi)^{-1}W_o(\theta')$.

The singly differential cross section c_o may be obtained as the
integral of the experimental $d^2 \sigma_o/dE'\theta'$, and according to an early
argument by Rudd and Macek[1], it may be related as $c_o = \sigma_c/(27.2 \text{ eV})$
to the reduced cross section σ_c of capture to Rydberg states. The results (o) for σ_c are presented in Fig. 5. Rødbro and Andersen[2] found
that c_o and σ_c could be derived from the peak value $\sigma(\epsilon)_{\theta_o}$ of the
forward-cone cusp if isotropy $W_o(\theta') = 1$ was assumed. In their Eq.
(A9), the factor 1.0005 should be corrected to 1.30, and the relation
obtained for $\theta_o = 0.38°$ and $\Delta E = 0.01E$ is

$$\sigma_c = 1.30\sigma(\epsilon)_{\theta_o} . \tag{6}$$

It is seen in Fig. 5 that the present forward-cone results (×) agree
with the results (·) obtained by Rødbro and Andersen and, furthermore,
that the forward-cone results agree quite well with the results (o)
of the small-angle measurements. This may indicate that Eq. (6) is
a a fair approximation, even with anisotropy in the rest frame of the
projectile.

Apart from the deviation of σ_c by the method developed by Rødbro and Andersen, the present analysis is based upon data from regions at such distances from the singularity that differential expe-

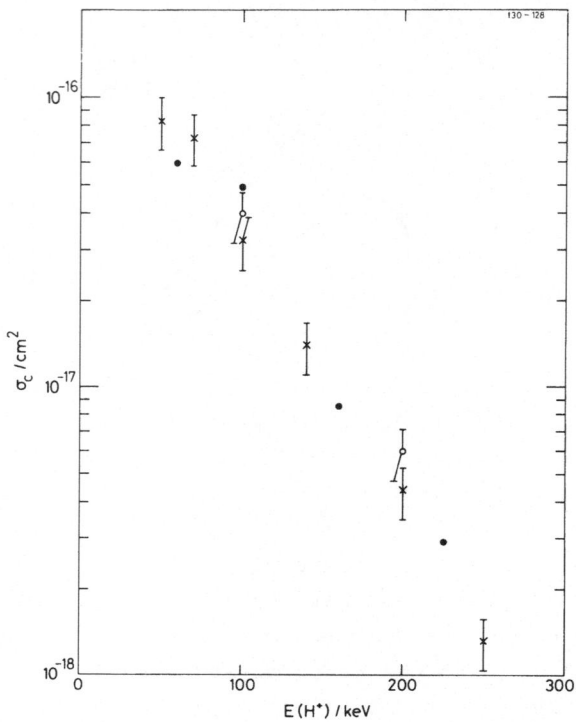

Fig. 5. Reduced cross sections σ$_c$ obtained from small-angle data (o) and from forward-cone data (×), together with forward-cone results from Rødbro and Andersen (•).

rimental conditions are obtained. This makes the data treatment simple but requires fine collimation in the experiments. It may be a useful alternative and supplement to the more commonly performed analysis[3] of the central part of the cusp spectrum.

References

1. M.E. Rudd and J. Macek, <u>Case Studies in Atomic Physics 3</u>, 47 (1972)
2. M. Rødbro and F.D. Andersen, J.Phys.B:Atom.Molec.Phys. <u>12</u>, 2883 (1979)
3. W. Meckbach, I.B. Nemirovsky, and C.R. Garibotti, Phys.Rev.A <u>24</u>, 1793 (1981)

THE INFLUENCE OF A DIFFUSE TARGET ON ELECTRON LOSS INTO THE CONTINUUM DOUBLE DIFFERENTIAL DISTRIBUTIONS

G.C. Bernardi, I.B. Nemirovsky, W. Meckbach, and C.R. Garibotti
Centro Atómico Bariloche - Comisión Nacional de Energía Atómica
8400 - Bariloche
Argentina

Measurements of double differential electron loss into the continuum (ELC) distributions in energy-or speed v-and angle θ of electrons ejected in H° → He collision have been recently reported (1). In that case electrons initially bound to the projectile are transferred to continuum states centered on the resulting ionic projectile.

If v_1 is the initial H° velocity and v' the electron velocity relative to the final proton, we have v'<< v_1 and a first Born perturbation treatment could be applied (2)(3)(4).

The apparatus and method of measurement were identical to those described in a preceding paper (5). To compare experiment and theory we had assumed that the electrons were originated at a point source given by a localized gas stream emerging from a hipodermic needle and the results for the H°→ He system at 105 keV beam energy were not in good agreement with the theoretical description of the ELC peak based in the first Born approximation (1). This leads us to a more careful analysis of the experimental setup.

The measured distribution showed an elongated shape-in v'-space-at small angles which suggest us that we were dealing with an extended gas target.

We checked the shape of the He target through some additional experiments. In these we compared the spectra obtained using the needle source with: a) Spectra from a uniform distribution of the gas at constant pressure inside the electron spectrometer, b) Spectra resulting after setting an electron suppressor electrostatic field on the beam path just behind the needle. From this comparision we conclude that an additional contribution of electrons came from the whole beam trajectory inside the inner cylinder of the spectrometer. To evaluate the extended electron source effect,a model was developed considering a row of point sources located along the beam path.The row is confined by the lenght of the proton beam in the spectrometer inner cylinder. The target gas density distribution in this zone can be approximately evaluated. Outside this limit the beam path -i.e. the electron sources- is exposed to the analyzer electric field.

Through a geometrical analysis the electron contribution along the extended collision region at each observation angle θ and angular acceptance θ_0 can be assessed.

To show the consequences of an extended source, we considered a theoretical $1/v'$ cross section and performed the convolution with the analyzer resolution. We found that the "experimental" electron distribution $Q'(v,\theta)$ resulting from the computed extended source can be related with that obtained from a point target $Q(v,\theta)$ by:

$$Q'(v,\theta) = Q(v,\theta) \ B(\theta) \qquad\qquad [1]$$

There is a simple factor $B(\theta)$, which depends on the angle θ as well as on the angular and velocity resolution of the equipment. It should be noted that it is independent of v and therefore previous studies performed on $\theta=0°$ spectra are not affected by the extended target correction.

In Fig. 1 we compare contour lines in v'-space. These lines correspond to constant values of the double differential distributions Q and Q' relative to the maximum at $v'=0$. It is seen that the lower level lines are the most affected by the extended target.

Data of electron experimental distributions from $H° \rightarrow H_e$ collision at 105 keV and $\theta_0 = 1.5°$ are shown in Fig. 2.

The evolution of the experimental peak height-taken at each peak velocity v_p -as a function of the emission angle θ is plotted in Fig. 3 together with the corresponding values of $Q(v,\theta)$(dashed line). The experimental and theoretical results have been normalized at $\theta=0°$.

The difference monotonically increases with θ.

The full line represents $Q'(v,\theta)$, i.e. $Q(v,\theta)$ multiplied by $B(\theta)$, Eq.[1]; good agreement with our experimental values is obtained.

Comparison between theory including anisotropic terms (4) (6) in $d\sigma/d\vec{v}$ and measured distribution including extended target corrections is in progress.

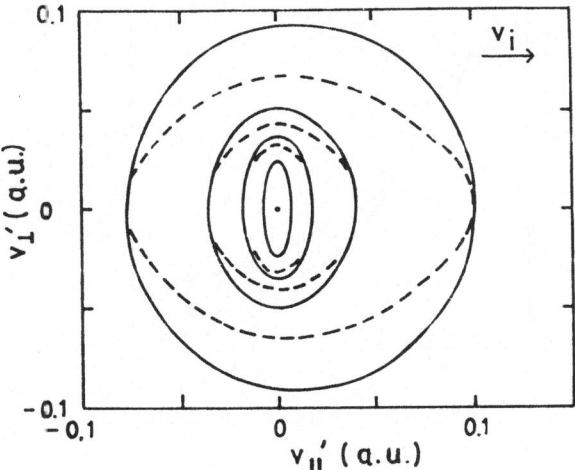

Fig. 1: Contour lines of Q(full lines) and Q'(dashed lines) distributions corresponding to a theoretical $1/v'$ cross section. Angular acceptance $\theta_0 = 1°$. Levels represent 0.2, 0.4, 0.6 and 0.8 fractions of the $v'=0$ peak height.

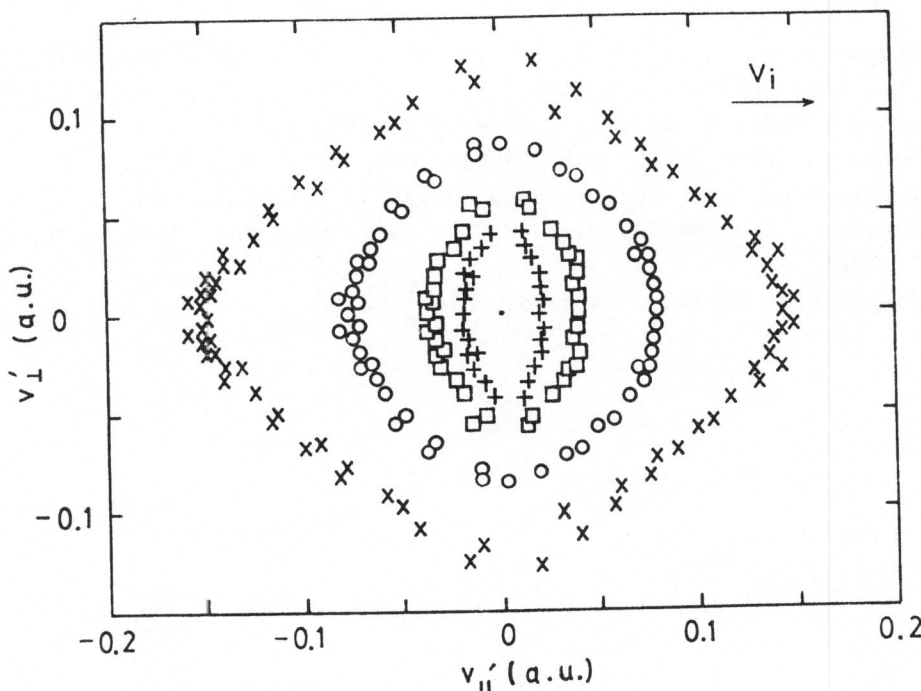

Fig. 2: Experimental contour lines for $H^\circ \rightarrow H_e$ collision at 105 keV beam energy and $\theta_0 = 1.5°$. Levels lines at : (x), 0.1; (O), 0.2; ($\square$), 0.4; (+), 0.6 of full peak height.

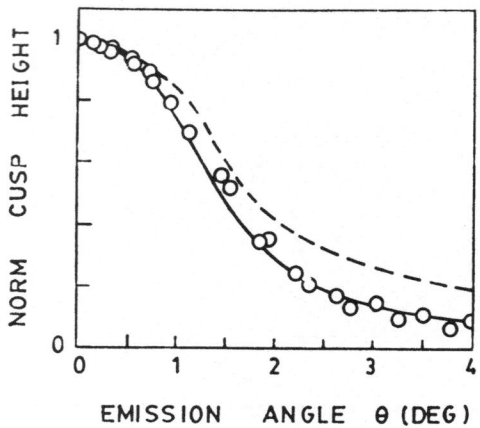

Fig. 3: Comparison of experimental (o) peak height variation with theoretical curves obtained from Q(---) and Q'(——) electron distributions.

REFERENCES

1. W. Meckbach, R. Vidal, P. Focke, I.B. Nemirovsky and E. Gonzalez Lepera, Phys.Rev.Lett. 52, 621 (1984).
2. F. Drepper and J.S. Brigss, J.Phys. B 9, 2063 (1976).
3. M.H. Day, J. Phys. B 13, L65 (1980).
4. J.S. Brigss and M.H. Day, J. Phys. B 13, 4797 (1980).
5. W. Meckbach, I.B. Nemirovsky and C.R. Garibotti, Phys. Rev. A 24, 1793 (1981).
6. J. Burgdörfer, Phys.Rev.Lett. 51, 374 (1983).

CUSP STUDIES FOR SIMPLE COLLISION SYSTEMS

D. Berényi, L. Gulyás, Á. Kövér, E. Szmola[x] and Gy. Szabó
Institute of Nuclear Research of the Hung. Acad.
Sci. /ATOMKI/, Debrecen, Pf. 51. Hungary, H-4001

M. Burkhard and K. O. Groeneveld
Institut für Kernphysik der J. W. Goethe Universität
6000 Frankfurt/M., F.R.G.

Since the discovery of the cusp-shaped peak[1] in the electron spectrum in forward direction from ion-atom collisions, at an energy at which the corresponding velocity of the electron is equal to the projectile velocity, a large number of papers have been published on this phenomenon. A detailed survey has been published recently[2].

In spite of the numerous works on the subject, some important information is missing in the case of simple collision systems as e.g. on the position, half-width /FWHM/, shape, intensity of the ECC and ELC peaks as a function of the impact energy. Such date are all the more interesting, because in the case of heavy-ion-atom collisions some deviations were observed from original theoretical predictions e.g. those on the impact energy dependence of the half width /FWHM/ of ELC cusps[3].

In the present work the features of the ECC and ELC cusps have been studied in case of the He^+, He^{++} — He simple collisions

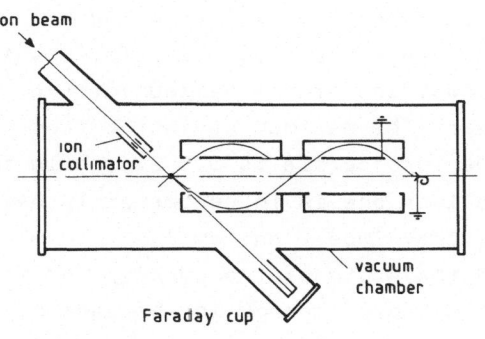

Fig. 1. Schematic diagram of the double "box" type electrostatic electron spectrometer[4].

Permanent address: Department of Physics, Technical University of Heavy Industry, Miskolc, Hungary

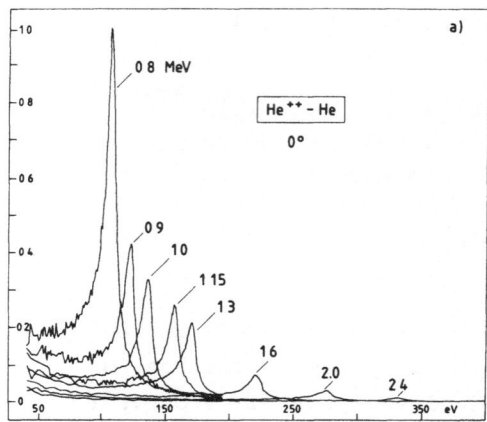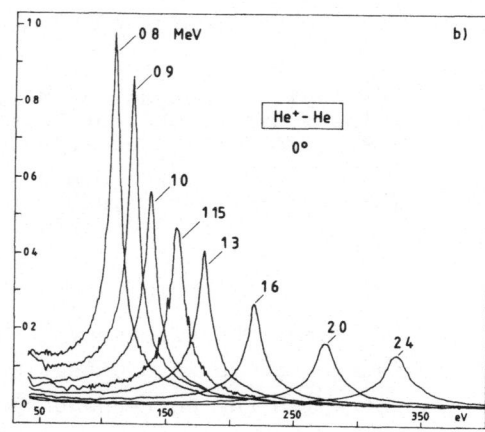

Fig. 2. The series of the cusps /DDCS at 0°/ for He[++] - He /a/ and He[+] - He /b/.

systems in the impact energy range from 0.8 to 2.4 MeV.

As an electron energy analyser a special double-pass /double "box" type — this "box" is similar to the cylindrical mirror spectrometer/ electrostatic electron spectrometer /ESA-13/ was used /Fig. 1/ made in ATOMKI, Debrecen[4/]. To set the required angle of the ejected electrons in the measurement, relative to the direction of the incident ions /here this required angle is 0°/, a proper cylinder was turned around inside the internal cylinder of the spectrometer. The angular acceptance of the spectrometer was ±1° and the energy resolution was 0.3 % in the present measurements. As a target, a gas beam was used with a nozzle of 1 mm diameter.

Fig. 2 shows the series of cusps for He[++] - He and He[+] - - He collision systems taken in the present study. The relationship between the impact energy and the position of the maximum of the cusp was found to be strictly linear. The maximum deviation from the straight line in the case of He[+] projectile is 0.5 eV, while in the case of He[++] it is 2.3 eV. The last one is found certainly because of the poor statistics at 2.4 MeV impact energy.

As can be seen in Fig. 3 the width of the cusps /FWHM/ varies linearly as a function of the impact energy in the case of He[++] projectile while for He[+] there is a deviation from the straight line predicted by the theory.

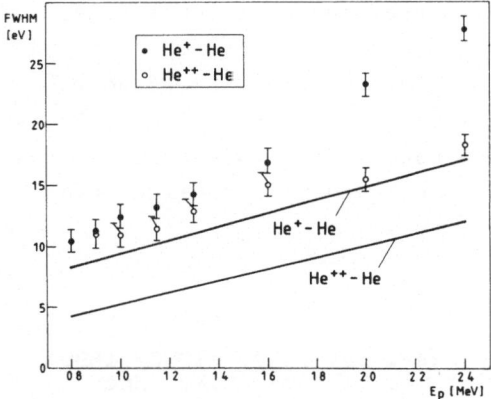

Fig. 3. Impact energy dependence of the width /FWHM/ of the cusps. Solid curves represent the results of the corresponding calculations /see the text/.

As regards the absolute values, there is a factor of about two disagreement between the experiment and the theory concerned. With increasing impact energy however, the cusp at He^{++} is more and more narrow relatively to that at He^+.

As regards the shape of the cusp, an approximate phenomenological description can be applied. The partial widths at the half maximum taken from the abscissa of the peak maximum towards lower /ΔE_-/ and higher /ΔE_+/ electron energies , respectively can be compared to characterize the asymmetry of the cusp. In the present study the $\Delta E_-/\Delta E_+$ ratio was found to be about 1 for He^+, and about 2 for He^{++} at every impact energy.

The variation of the cusp intensity as a function of the impact energy is given in Fig. 4. The general course of the vari-

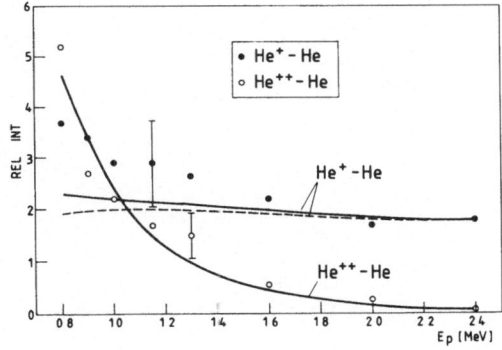

Fig. 4. Variation of the intensity of the cusps as a function of the impact energy. The curves are the results of the corresponding calculations /see the text/. In case of the dashed curve, the excitation of the target atom is taken into account.

74

ation is described rather well by the corresponding calculations. The curves are normalized at 2.4 MeV.

The calculations for He^{++} were carried out according to Salin[5] by using hydrogenic wavefunctions. In the case of He^{+}, Roothaan-Hartree-Fock wavefunctions[6] were used in the PWBA calculations and the excitation of the target atom was taken into consideration according to Day[7].

References

1/ G.B. Crooks and M.E. Rudd, Phys. Rev. Lett., 25 /1970/ 1599
2/ M. Breinig, S.B. Elston, S. Huldt, L. Liljeby, C.R. Vane, S.D. Berry, G.A. Glass, M. Schauer, I.A. Sellin, G.D. Alton, S. Datz, S. Overbury, R. Laubert and M. Suter, Phys. Rev. A 25 /1982/ 3015
3/ M. Breinig, M.M. Schauer, I.A. Sellin, S.B. Elston, C.R. Vane, R.S. Thoe and M. Suter, J. Phys. B: At. Mol. Phys. 14 /1981/ L291
4/ A. Kövér, D. Varga, E. Szmola, Gy. Mórik and J. Herbák, to be published in Nucl. Instr. Meth.
5/ A. Salin, J. Phys. B: At. Mol. Phys., 5 /1974/ 979
6/ E. Clementi, C. Roetti, Atomic Data and Nuclear Data Tables 14 /1974/ 177
7/ M.H. Day, J. Phys. B: At. Mol. Phys. 14 /1981/ 231

DOUBLY DIFFERENTIAL EMISSION DISTRIBUTIONS
FOR ELECTRON LOSS TO THE CONTINUUM
FROM FAST HEAVY PROJECTILES IN GAS TARGETS*

S. B. Elston, S. D. Berry, M. Breinig, R. DeSerio,
C. E. Gonzalez Lepera, and I. A. Sellin
Dept. of Physics, University of Tennessee, Knowville, TN 37996
and
Physics Division, Oak Ridge National Laboratory, Oak Ridge, TN 37831

K.-O. Groeneveld, D. Hofmann, and P. Koschar
Institut fur Kernphysik der Universitat Frankfurt/M
Federal Republic of Germany

I. B. Nemirovsky
Centro Atomico Bariloche and Instituto Balseiro
8400 Bariloche, Argentina

L. I. Liljeby
Research Institute of Physics, Stolkholm, Sweden

We have constructed an apparatus which utilizes position-sensitive detector technology for the measurement of doubly differential electron emission cross-sections in ion-atom and ion solid collisions at emission angles near the forward direction. The properties of the instrument make it especially well suited for the simultaneous doubly differential measurement of electron emission processes which are centered in the frame of the projectile, and so it is of value for investigations of the well known electron capture and loss to projectile-centered continuum state processes (ECC and ELC) which have received much attention at this symposium. Further, the multichannel nature of the system leads to improved data acquisition efficiency, sometimes an important advantage for the observation of processes which require the use of heavily-subscribed accelerator user facilities. We describe here results of some of the first measurements to be made with this apparatus, specifically the doubly differential (in velocity and angle) cross section (DDCS) for continuum transfer emission from collisions of 41, 82, and 105 MeV O^{5+} in helium and argon gas targets. Because the projectiles carry a relatively loosely bound 2s electron into the collision, ELC processes are expected to dominate the emission characteristics.

The full-width-at-half-maximum (FWHM) of the cusp formed by ELC for high-Z ions has been reported[1] to be smaller than expected and to not scale proportionately with the velocity of the projectile, as predicted by the earliest theories[2] of the process. While later theories[3] explicitly provided for anisotropic emission in the frame of the projectile which in principle could account for such behavior of the singly differential cusp spectra as observed in the laboratory frame, details of the observations could not be quantitatively accounted for. For example, highly charged projectiles such as O^{5+} and C^{4+} with relatively loosely bound L-shell electrons were found to have narrow widths ~0.25 - 0.3 a.u. (in emission velocity) almost independent of target, projectile velocity, or projectile charge over a range of collision velocities between 6 and 12 a.u. Most calculations treated atomic hydrogen targets only and assumed tightly bound 1s projectile electrons available for the loss process. A recently reported calculation by Burgdörfer et al.[4] has made available some relevant predictions and presents a method that promises detailed numerical results for comparison with experimentally practical collision systems.

A central feature of the calculation performed by Burgdörfer is that the narrow ELC cusps, with collision velocity-independent widths, observed for high-Z ions over a wide range of collision velocities are a consequence of preferential transverse electron emission from the 2s level in 3- and 4-electron ions coupled with convolution over the narrow range of observation angles admitted in most ELC experiments. If the DDCS in the projectile frame is expanded in multipoles as is usual for anisotropic emission, it can be parameterized as

$$d\sigma/d\vec{v} = a_0\left[1 + \beta_2 P_2(\cos\theta) + \beta_4 P_4(\cos\theta)\right],$$

where the emission velocity v and polar angle θ are expressed in the projectile frame, P_2 and P_4 are Legendrè polynomials, a_0 sets the isotropic emission level, and the second and fourth order coefficients β_2 and β_4 determine the degree and nature of the anisotropic component of emission. Various symmetry considerations prohibit other multipoles in the expansion of the DDCS for pure ELC processes. For the projeciles, targets, and velocities studied here, $\beta_2 \sim -0.6$ and $\beta_4 \sim +0.1$, which leads to an emission pattern which is strongly transverse to the ion beam. Our preliminary measurements of the emission distribution, seen in Figure 1, exhibit definite transverse anisotropy.

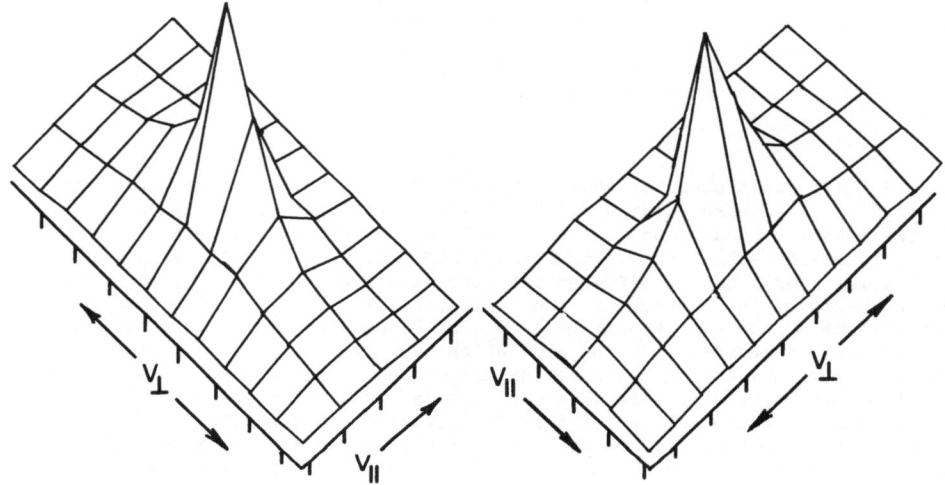

Figure 1. Two views of the DDCS as seen in projectile frame for 82 MeV O^{5+} in He. Transverse ($v_\perp$) and longitudinal ($v_\parallel$) velocity axes are graduated in units of 0.1 a.u.

In Figure 2, we compare plots of contours of equal intensity as observed in the laboratory frame for 82 Mev O^{5+} on He and Ar targets with the results of a simulation based on the asymmetry parameters quoted above and upon a convolution with the instrumental response function we expect from essentially geometric considerations. Corrections for transformation between projectile and laboratory frame and spectrometer transmission efficiency are also included. Note that fewer contours are displayed for the simulated data. One can see that the degree of asymmetry exhibited by the data away from the central portion of the cusp (where the singular nature of the cusp results in extreme sensitivity to the details of the instrumental response function) is reasonably good. The cusps obtained with argon targets are a bit narrower than the simulated data, and the helium cusps are significantly narrower overall than expected, a puzzling result if one expects the calculations to be more reliable for the simpler target.

The major elements of the apparatus are diagrammed in Figure 3. The target region, which is a ~0.5 cm thick gas cell in the present measurements, is viewed by a spherical sector electrostatic spectrometer having a mean deflection radius of 5.5 cm and a deflection angle of 160 deg. An important feature of this spectrometer design is that, to first order and in the absence of extraneous fields, it provides

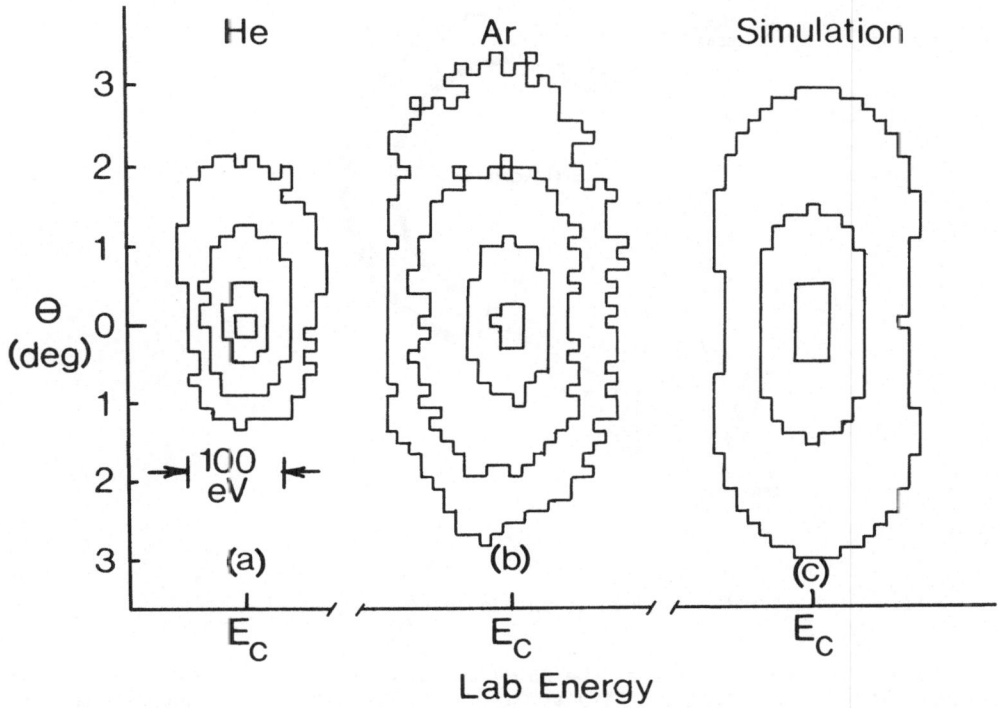

Figure 2. Contour plots of the convoluted DDCS for (a) helium and
(b) argon targets, and (c) the simulated data. Contours plotted
are 0.5, 0.2, 0.1, and 0.05 times the peak value, working outward
from the center. The last contour is not plotted in (c).

focussing in both the deflection plane and in the plane normal to it. Thus, for
electrons emitted within the target volume at a point conjugate to the exit aperture
of the spectrometer (1.0 mm diameter in the present case), the emission angles
correspond one-for-one with arrival angles at the exit aperture. Thus, by including a
short drift region between the exit aperture and a position-sensitive charged
particle detector, the preserved emission angles can be converted to detector
position coordinates. In our apparatus, the drift space is about 15 cm in length and
the detector consists of a tandem chevronned pair of microchannel plate electron
multipliers, which preserve the position information of the primary event during the
amplification process, followed by a circular arc terminated resistive anode of the
kind first described by Lampton and Carlson[5]. Four charge pulse outputs of the
anode are independently amplified and then decoded according to a simple ratiometic
method. While the intinsic positional resolution of the detector itself is 100 to 150
μm, the angular resolution provided by the apparatus is determined primarily by the
angle subtended by the exit aperture as seen from the detector entrance. In the
present case, the angular resoluion is about 6.7 mrad (0.38 deg.) and the
spectrometer energy resolution is 0.9% (FWHM). Not shown in the diagram are a pair of
parallel mesh grids (80% transparent) oriented perpendicular to the axis cf the drift
space, spaced 3 mm apart, and located immediately prior to the entrance to the
microchanel plates, which are used to reject stray, low energy electrons. Also not
shown are a set of transverse deflection plates which can be used both to compensate
for deflection of analyzed electrons resulting from residal stray magnetic fields and
as diagnostic elements. The local magnetic field in the region of the spectrometer is
nulled by a triplet of coils external to the vacuum system. Target gas pressures are
controlled by a piezoelectic valve and monitored by a capacitance manometer.

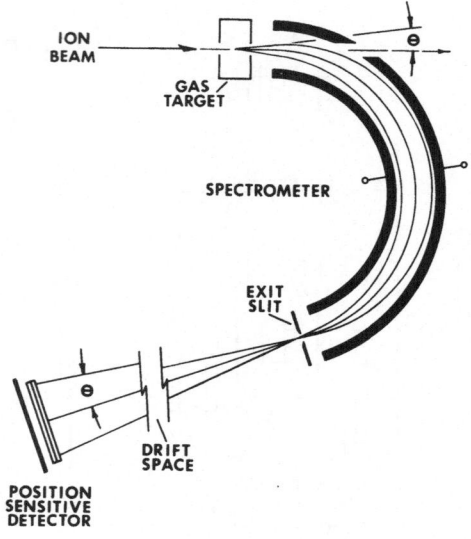

Figure 3. Schematic diagram of the apparatus.

The data were collected at the 25 MV tandem electrostatic accelerator of the Holifield Heavy Ion Research Facility at Oak Ridge; the authors are indebted to the staff of that facility for their services.

*Work supported in part by NSF, Physics Division and Division of International Programs, and in part by the U. S. Department of Energy, Office of Basic Energy Sciences under Contract No. DE-AC05-84OR21400 with Martin Marietta Energy Systems, Inc.

[1]M. Breinig, M. M. Schauer, I. A. Sellin, S. B. Elston, C. R. Vane, R. S. Thoe, and M. Suter, J. Phys. B 14, L291 (1981). For a more comprehensive review, see also M. Breinig, S. B. Elston, S. Huldt, L. Liljeby, C. R. Vane, S. D. Berry, G. A. Glass, M. Schauer, I. A. Sellin, G. D. Alton, S. Datz, S. Overbury, R. Laubert, and M. Suter, Phys. Rev. A 25, 3015 (1982).

[2]F. Drepper and J. S. Briggs, J. Phys. B 9, 2063 (1976).

[3]J. S. Briggs and F. Drepper, J. Phys. B 11, 4033 (1978); J. S. Briggs and M. H. Day, ibid. 13, 4797 (1980); M. H. Day, ibid. 14, L65 (1980); 14, 231 (1981).

[4]J. Burgdörfer, M. Breinig, S. B. Elston, and I. A. Sellin, Phys. Rev. A 28, 3277 (1983).

[5]M. Lampton and C. W. Carlson, Rev. Sci. Instr. 50, 1093 (1979).

PROJECTILE CONTINUUM ELECTRONS
IN HIGHLY CHARGED ION-ATOM COLLISIONS

Lars H. Andersen
Institute of Physics, University of Aarhus
DK-8000 Aarhus C, Denmark

Abstract

The emission of electrons at zero degree in collisions of 20-MeV Au^{q+} ($5 \leq q \leq 19$) with He has been measured in coincidence with final projectile charge state. Autoionization lines are observed, which are due to transfer and excitation. The large transfer-ionization cross section observed is accounted for by transfer of two electrons to a highly correlated ion state, followed by loss of one electron to the continuum. The ionization cross section is entirely accounted for by electron capture to the continuum for $q \gtrsim 10$.

Introduction

The collision between highly charged, partially stripped ions at intermediate velocity $V \simeq v_0$ (v_0 being the Bohr velocity) offers a good opportunity to study several fundamental collisional processes such as electron transfer, ionization and excitation, and combinations of these processes. A number of coincidence measurements of electron capture and target ionization for such collision systems have recently been reported[1-4]. In this paper, we consider the following collision process, for which the partial cross section $\sigma_{q,q'}^{on}$, have been measured[1],

$$(20\text{-MeV}) \quad Au^{q+} + He^{\circ} \rightarrow Au^{q'+} + He^{n+} .$$

It was observed that the number of electrons captured by the projectile does not necessarily equal the number of electrons released from the target. The term 'transfer ionization' (TI) is used for such processes, where the number of electrons released from the target exceeds the number captured by the projectile. Normally, the partial cross sections $\sigma_{q,q'}^{on}$, are obtained by charge-state analyzing the projectile and the target atom after the collision. By doing so, no information is obtained about electrons released in the collision. In the present work, energy spectra of the ejected electrons have been measured in the forward direction. From the experimental data, information about electron capture to the continuum (ECC), simultaneous electron transfer and excitation (TE), and transfer ionization (TI) was obtained since these processes lead to electrons that are slow in the rest frame of the ion. Due to the small scattering angle in the laboratory frame, these electrons are efficiently detected at zero degree.

The partial cross sections $\sigma_{q,q'}^{on}$, for 20-MeV Au^{q+} + He measured by Damsgaard et al.[1] are shown in Fig. 1.

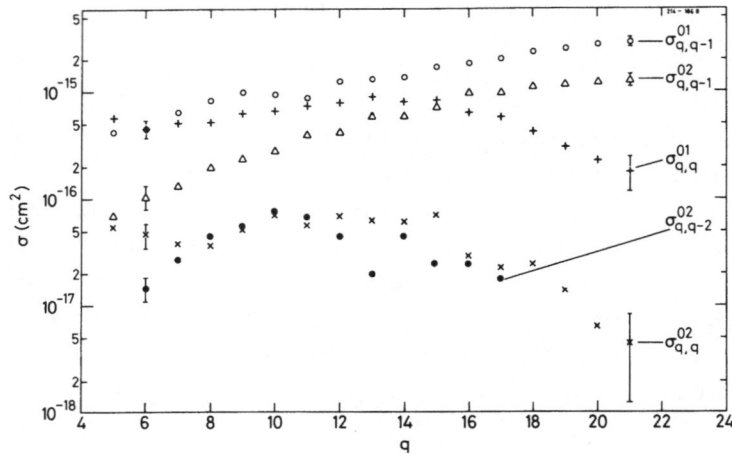

Fig. 1. Partial cross sections $\sigma_{q,q'}^{on}$, as a function of q for 20-MeV Au^{q+}+He.

From the experimental data, we observe that the transfer-ionization cross section $\sigma_{q,q-1}^{02}$ approaches half the single-capture cross section at q $\gtrsim$ 13. Further, it is noted that the double-capture cross section $\sigma_{q,q-2}^{02}$ is one or two orders of magnitude lower than the transfer-ionization cross section. We have recently shown[5] that this is due to the fact that when two electrons are captured by the highly charged projectile, they enter a highly correlated state which autoionizes with a probability close to unity. The correlated two-electron state causes the electrons to be emitted in a continuous rather than in a discrete manner[5].

In 1978, is was demonstrated by Shakeshaft[6] that electron capture to the continuum contributes about 50% to the total-ionization cross section for H$^+$ + H at ~40 keV. For the present collision system, we find that close to 100% of the ionization cross section $\sigma_{q,q}^{01}$ can be accounted for by ECC for q $\gtrsim$ 10.

Experimental Technique

The 20-MeV gold beam was provided by an EN tandem accelerator. A poststripper facility (carbon foil or gas) produced ions of the desired charge state, which were selected by magnetic deflection. The beam was electrostatically cleaned immediately before the target. After the target, the beam was charge-state-analyzed, and electron-energy spectra

were recorded at zero degree, in coincidence with a given exit charge
state. The spectrometer used in this work was a 30° parallel-plate analyzer[7].

The experimental data are normally presented as the laboratory
cross section as a function of the laboratory kinetic electron energy.
However, useful information was obtained when the electron-energy spectra were transformed into the rest frame of the projectile. Here the
projectile rest frame energy E_P is given by

$$E_P = E_L + RIE - 2(RIE \cdot E_L)^{\frac{1}{2}} \cos\theta , \tag{1}$$

where E_L is the laboratory kinetic energy, RIE is the reduced ion ener-
gy $\frac{1}{2}m_e V^2_{ion}$, and θ is the laboratory scattering angle. At zero degree
($\theta = 0°$),

$$E_P = (\sqrt{E_L} - \sqrt{RIE})^2 . \tag{2}$$

Due to the finite acceptance angle of the spectrometer, corrections
to Eq. (2) were taken into account. Let $E_P^* \equiv (\sqrt{E_L} - \sqrt{RIE})^2$ and (φ, ψ) be
the scattering angles in the rest frame of the projectile perpendicular
to the beam direction. To first order, we then have

$$E_P(\varphi, \psi) = E_P^*(tg^2\varphi + tg^2\psi + 1) . \tag{3}$$

The corrected E_P was obtained by integrating $E_P(\varphi, \psi)$ over φ and ψ, assuming isotropic emisison.

The double-differential cross section in the rest frame of the
projectile is given by

$$(\frac{d^2\sigma}{dEd\Omega})_P = (\frac{d^2\sigma}{dEd\Omega})_L (\frac{E_P}{E_L})^{\frac{1}{2}} . \tag{4}$$

If the cross section in the rest frame of the projectile is a slowly
varying function of E_P, it is seen from Eq. (4) that the laboratory
cross section has a singularity at $E_P = 0$. This singularity gives rise
to the 'cusp'.

Results and Discussion

Electron-energy spectra were measured in the forward direction for
20-MeV Au^{q+} ($5 \leq q \leq 19$) on He and for 20-MeV Au^{15+} on H$_2$ and Ar. The spectra were dominated by the cusp, which is due to transfer to the continuum processes.
In addition to the cusp, some autoionization lines were observed on the
wings of the cusp. First, we discuss the line emisssion.

Several mechanisms could lead to creation of doubly excited states
which, in turn, decay by electron emission through an autoionization
process. Considering only the simplest processes, the doubly excited
states could be formed by (i) double-core excitation,

$$Au^{q+} + T \rightarrow Au^{q+**} + T \rightarrow Au^{(q+1)+} + T + e^- \ ,$$

(ii) single capture to an excited state plus core excitation,

$$Au^{q+} + T \rightarrow Au^{(q-1)+**} + T^+ \rightarrow Au^{q+} + T^+ + e^- \ ,$$

and (iii) double-electron capture to excited states,

$$Au^{q+} + T \rightarrow Au^{(q-2)+**} + T^{2+} \rightarrow Au^{(q-1)+} + T^{2+} + e^- \ ,$$

(the symbol T is used for any one of the targets). To reveal the origin of the autoionization process, experiments were performed where the electron-energy spectra were measured in coincidence with the final charge state of the projectile. Figure 2 shows an example of such coincidence spectra for q = 17+ and helium as target.

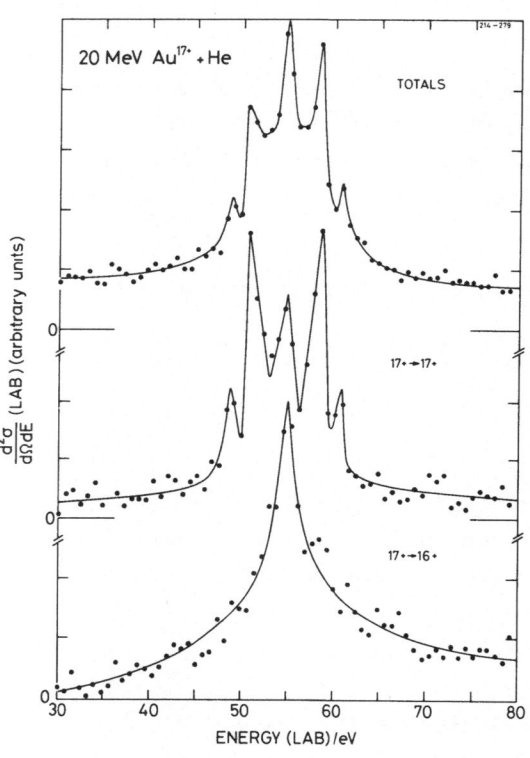

Fig. 2. Double-differential cross section in the laboratory frame as a function of electron energy for 20-MeV Au^{17+} on He. Counts at cusp centers are 886, 79, and 125 for the 'totals' 17+ → 17+ and 17+ → 16+ spectra, respectively. Lines are drawn to guide the eye.

The lines are clearly associated with the 17+ → 17+ coincidence channel[5], and according to (ii), they are due to electron transfer and projectile-core excitation (TE). The cusp is composed of contributions from ECC (17+ → 17+) and TI (17+ → 16+).

The simultaneous electron capture and excitation is qualitatively analogous to dielectronic recombination (inverse Auger transition). Therefore the process may have a resonant characteristic. Tanis et al.[8]

found a resonant behaviour in the projectile x-ray yield associated with electron capture in S+Ar collisions; however, a close association to dielectric recombination remains to be established. The possibility that resonant transfer and excitation (RTE)[9] is playing a rôle for our collision system exists, but its relative importance has not yet been evaluated.

The cross section for TE followed by electron emission for Au^{15+} + He is found to be $\sim 5 \times 10^{-17}$ cm², which is much less than the single-capture cross section (= 1.7×10^{-15} cm²) at this energy (Damsgaard et al.[1]). Since the geometrical size of the ion core is $\simeq \pi a_0^2 = 8 \times 10^{-17}$ cm², we expect the TE process to be controlled by the core excitation. That is, at impact parameters leading to core excitations, the probability for capture is close to unity. According to recent calculations of Mc-Dowell and Janev[10], the most strongly populated level in the capture process for Au^{15+} on He is n = 6. This level is associated with an ionization potential of ~ 80 eV. Thus the excitation in question must be of the same order of magnitude. Preliminary relativistic Dirac-Fock calculations[11] show that the 4f → 5p core-excitation energy of Au^{15+} is indeed $\sim 60-85$ eV.

The target dependence of the autoionization process was investigated by comparing coincidence spectra (q+ → q+) of 20-MeV Au^{15+} on H_2, Ar, and He. In Fig. 3, the result of this comparison is shown. We find the energies and the relative intensities of the lines to be independent of the target species. Thus the doubly excited states are specific for the ion core and do not depend on the target. The cusp seen on the three spectra is due to ECC. The cusp yield is relatively low for the helium target, due to the high ionization potential for helium (24.6 eV compared to 15.8 for Ar and 15.4 for H_2). This causes the lines to appear much larger in this case.

We now turn to the continuous part of the electron-energy spectra. Evidently, the cusp from a noncoincidence measurement is composed of several contributions. For the lower-charge states, the cross section for projectile ionization is relatively high, and the cusp will contain contributions from ELC as well as from ECC and TI (as mentioned earlier, TI is the process where the number of electrons released from the target exceeds the number captured to bound states on the projectile). For the higher charge states, ELC is unlikely, and the cusp is mainly composed of contributions from ECC and TI.

To characterize the cusps measured in the coincidence channels, q+ → q+ (ECC) and q+ → (q-1)+ (TI), we have applied an expansion method, in which the general scattering amplitude of the transfer to the con-

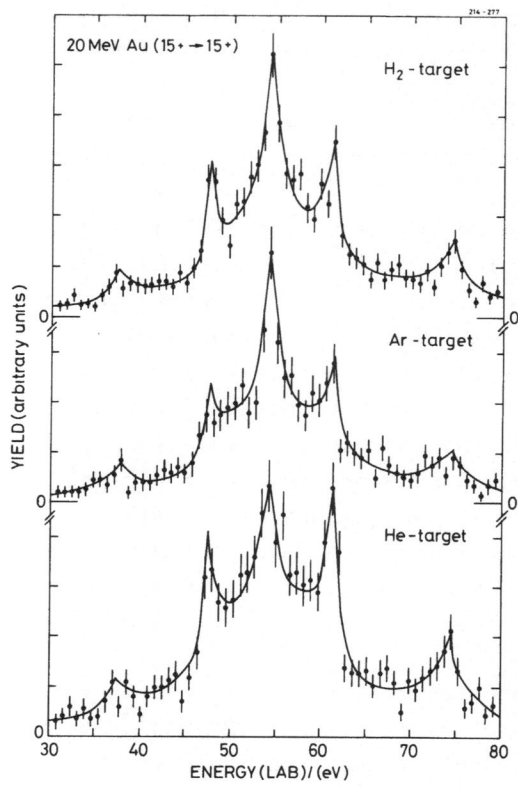

within figure: 20 MeV Au (15+ → 15+), H₂-target, Ar-target, He-target, YIELD (arbitrary units), ENERGY (LAB)/(eV), 30 40 50 60 70 80, 214-277

Fig. 3. Coincidence – electron-energy spectra of electrons emitted in the forward direction for 20-MeV Au (15+ → 15+) on H_2, Ar, and He targets. Lines are drawn to guide the eye. The data are not corrected for the transmission efficiency of the spectrometer.

tinuum process is partial-wave expanded. The method has been used recently to characterize cusp shapes for a number of collision systems (Refs. 12-16). In the rest frame of the projectile, we obtain

$$(\frac{d^2\sigma}{d\Omega dE})_p = \sum_{\ell} B_\ell P_\ell (\cos\theta')$$

$$= B_0^{(0)} + B_1^{(0)}\cos\theta' + V_p B_0^{(1)} + V_p B_1^{(1)}\cos\theta' , \qquad (5)$$

and in the laboratory frame (see Eq. (4))

$$(\frac{d^2\sigma}{d\Omega dE})_L = V_L (B_0^{(0)}/V_p + (B_1^{(0)}/V_p)\cos\theta' + B_0^{(1)} + B_1^{(1)}\cos\theta') . \qquad (6)$$

In Eq. (5), two Legendre polynomials and two terms in the Taylor expansion of B_ℓ are included (V_L and V_p are the laboratory and projectile -electron velocity in a.u., respectively, $B_\ell^{(j)}$ are the expansion coefficients, and θ' is the scattering angle in the rest frame of the protile). The singular character of the cross section is represented by the terms $B_0^{(0)}$ and $B_1^{(0)}$, whereas $B_0^{(1)}$ and $B_1^{(1)}$ show nonsingular behaviour. Further, $B_0^{(0)}$ and $B_0^{(1)}$ have an s-wave character, while the two terms $B_1^{(0)}\cos\theta'$ and $B_1^{(1)}\cos\theta'$ are characteristic of a p wave. These latter terms represent nonsymmetric terms in the expansion and have been interpreted as being contributions from second-order Born terms (Refs. 17,18). By integrating the differential cross section (Eq. (6)), including the spectrometer transmission function, over the experimental acceptances in velocity and angle, an expression is obtained, which can be fitted to the experimental data.

To examine the ECC cusp, we performed such a fit with the 20-MeV Au^{11+} + He → Au^{11+} coincidence electron-energy spectrum since in that particular case, no autoionization lines were observed. When fitting with s waves alone, we obtained $x^2 = 3.3$, whereas the inclusion of p waves reduces x^2 to 1.3; hence, in order to obtain a satisfactory fit to such an ECC cusp, p-wave components must be included. The coefficients obtained are listed in Table I. Also listed are the coefficients

Table I

Coefficients $B_i^{(j)}$ obtained from fitting to experimental coincidence cusps for 20-MeV Au^{q+} + He.

		11+ → 11+	15+ → 14+
s waves	$B_0^{(0)}$	745 ±15	611 ±15
	$\dfrac{B_0^{(1)}}{B_0^{(0)}}$	0.08± 0.08	1.0± 0.1
p waves	$\dfrac{B_1^{(0)}}{B_0^{(0)}}$	-0.23± 0.04	0
	$\dfrac{B_1^{(1)}}{B_0^{(0)}}$	-0.38± 0.18	0
	x^2	1.3	1.3

obtained by fitting to the Au^{15+} → Au^{14+} coincidence cusp. In this case, no significant decrease in x^2 was obtained by including p waves. The data for the ECC and TI cusps, together with the best fit (yielding the values given in Table I) are shown in Fig. 4. Evidently, the partial-wave expansion gives a good fit to the data. On top of Fig. 4, the results of the ECC and TI fits are compared (the yield is set equal at the cusp peak). The difference in cusp shape is obvious; the ECC cusp has a much lower yield at the high-energy side of the cusp relative to the low-energy side than has the TI cusp. The difference becomes even more pronounced when the data are transformed into the rest frame of the projectile. This is demonstrated in Fig. 5, where the ECC cusp

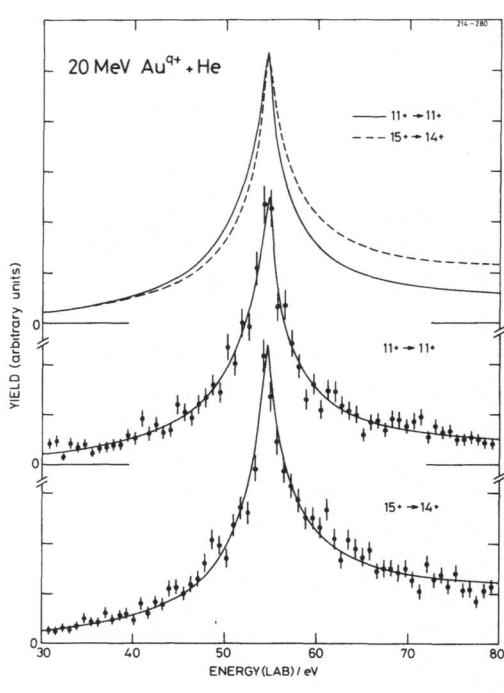

Fig. 4. *Coincidence -electron-energy spectra for 20-MeV Au^{q+} + He. Shown are the two spectra 11+ → 11+ (ECC) and 15+ → 14+ (TI) together with the best fit obtained with the partial-wave expansion method. On top of the figure, the two fits are compared. The data are not corrected for the transmission efficiency of the spectrometer.*

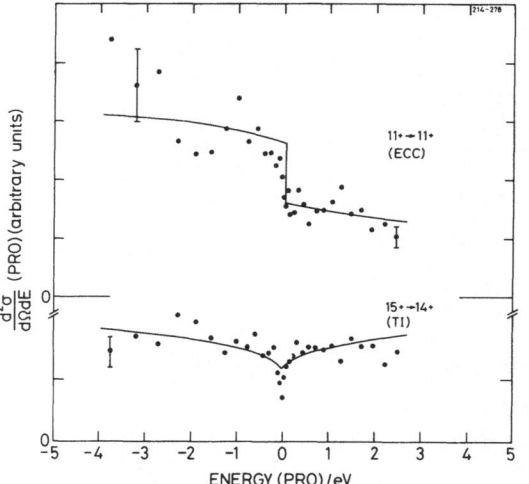

Fig. 5. *Double-differential cross sections in the rest frame of the projectile as a function of electron energy. Shown are the experimental data for 11+ → 11+ and 15+ → 14+. Shown are also curves calculated from the expression given in Eq. (5), with the fit parameters listed in Table I.*

(11+ → 11+) and the TI cusp (15+ → 14+) are shown. Also the expression (5) with the coefficients obtained in Table I for the two cusps is shown. The TI cusp is found to be symmetric with respect to zero energy. This is characteristic for the s-wave contributions. The ECC cusp, on the other hand, shows a strong asymmetric behaviour with a discontinuity at zero energy[17]. This drop is determined by the value of $B_1^{(0)}$, which is found with relatively good accuracy from the fit (see Table I). The slope of the curves is determined by $B_0^{(1)}$ and $B_1^{(1)}$.

From the study of the cusp shapes for q = 15, we find that the TI cusp does not exhibit an enhanced yield of electrons on the low-energy side of the cusp, which is characteristic for the ECC cusp[19]. On the contrary, the cusp shape is similar to that of an ELC cusp[19,20]. We find similar q → q-1 coincidence cusps for the other targets (H_2, Ar).

The mechanism for TI clearly involves two electrons. Based on the shape analysis, we suggest that both electrons are transferred to the rest frame of the moving ion, whereupon one is lost to the continuum and the other is left in a bound state. We suggest this to happen because the two target electrons enter a projectile state, which is highly correlated: The size of the two-electron wave packet is close to that of the helium atom, which is much smaller than the extension of the wave function for two corresponding but uncorrelated electrons on the projectile. The spatially correlated state autoionizes with a probability close to unity, and as a result of the short lifetime, the electrons are emitted with a continuous energy spectrum, in contrast to what is expected for electrons emitted from an uncorrelated state. With helium as the target, we find that the cross section in the q → q-1 cusp equals $\sigma_{q,q-1}^{02}$, the TI cross section. With the present model, we can account for the TI cusp as well as for the magnitude of the double -capture cross section $\sigma_{q,q-2}^{02}$, which is much smaller than $\sigma_{q,q-1}^{02}$. We would like to emphasize that at present, no detailed calculation exists which can elucidate the details of the two-electron correlation effect observed in these collisions.

The total TI cross section $\sigma_{q,q-1}^{02}$ for 20-MeV Au^{q+} + He has recently been calculated by McDowell and Janev[10]. They used the independent-particle model, in which the probability for simultaneous capture and ionization at a given impact parameter P_{TI} is given by the product $2P_c \cdot P_i$, where P_c is the probability for capture, and P_i is the probability for ionization. The total cross sections were obtained by using the classical-trajectory Monte-Carlo method. In this model, no correlation effects are taken into account. In Fig. 6, the result of these calculations are compared with the experimental transfer-ionization cross section $\sigma_{q,q-1}^{02}$. For q ≲ 10, the predictions of the independent-particle model are in good agreement with the experimental data. For the higher charge states of the projectile, however, the calculated cross sections are much too small. In Fig. 6, we also show the result of a 'potential-barrier model'[10]. In this model, TI is accounted for in terms of transfer of two electrons to a doubly excited state, which autoionizes with

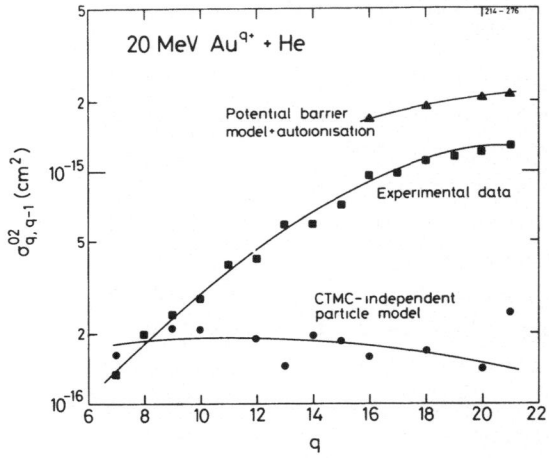

Fig. 6. Transfer-ionization cross section as a function of q for 20-MeV Au^{q+} + He. Shown are the experimental data (Ref. 1) and two theoretical approaches of McDowell and Janev (Ref. 10). Lines are drawn to guide the eye.

the probability ω. For ω, we take the ratio $\sigma^{02}_{q,q-1}/(\sigma^{02}_{q,q-1}+\sigma^{02}_{q,q-2})$ from the experimental data[1]. For the higher charge states, $\omega \gtrsim 95\%$. Evidently, these simple calculations support the double capture + autoionization mechanism for higher q values. If, as discussed above, the autoionization process gives rise to a continuum of slow electrons in the projectile rest frame, this model also accounts for the observed TI cusp. For $q \leq 10$, the calculations predict that the TI cross section should be accounted for in terms of independent capture + ionization. Since ionization basically results in ECC (see discussion below), the TI mechanism should be accounted for in terms of simultaneous bound-state and continuum capture. For $q \lesssim 11$, we do indeed find that the $q \rightarrow q-1$ coincidence cusp contains some p-wave character, indicating that at the low q values, the TI cusp contains electrons which are captured to the continuum (ionized), coincident with single-capture to bound states[21].

We shall now proceed with a discussion of the total cross sections obtained from the cusp yield (only for helium target). To convert the cusp yield into absolute cross sections, we applied the normalization procedure previously used by Vane et al.[21]. In our case, the ECL-coincidence cusp yield was measured for Au^{7+} + He $\rightarrow$ Au^{8+} + He + e^-. This yield, in turn, was normalized to the total electron-loss cross section obtained from independent measurements[22].

In order to discuss ECC, which is a capture-type process, but which leads to target ionization (σ^{01}_{qq}), we applied a simple approach based on a classical model of interaction[23,24]. Within this model, target ionization is possible only for the lowest values of q for our collision system. Experimentally, we find that ECC can account for $\sigma^{01}_{q,q}$ at $q \gtrsim 10$, but at $q \lesssim 10$, the yield in the ECC cusp cannot account for the observed

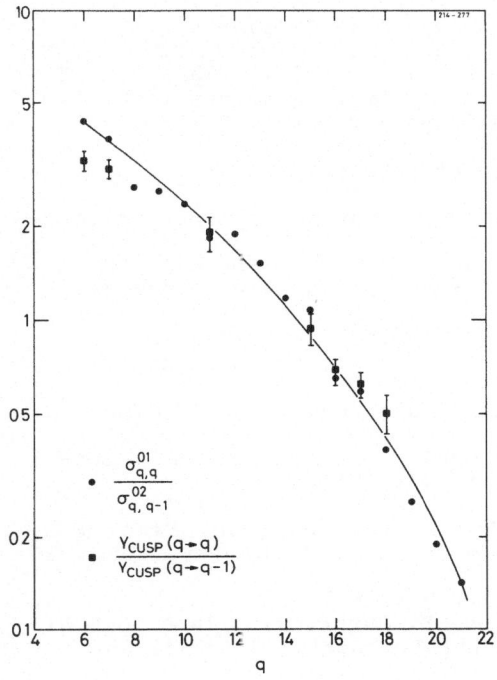

Fig. 7. The ratio of the cusp yield in the two coincidence channels q → q and q → q-1 is compared with the experimentally obtained total cross-section ratio $\sigma_{q,q}^{01}/\sigma_{q,q-1}^{02}$ (Ref. 1). A curve is drawn to guide the eye.

ionization cross section, in agreement with the simple classical model, which predicts 'true' target ionization to take place.

We also find the yield in the q → q-1 coincidence cusp to account for the TI cross section $\sigma_{q,q-1}^{02}$. When considering the ratio of the cusp yield in the two channels q → q and q → q-q, we find a value which should equal $\sigma_{q,q}^{01}/\sigma_{q,q-1}^{02}$, independent of the normalization method applied. As demonstrated in Fig. 7, this holds for all q values (q ≤ 18) except for q = 5 and 6, where some 'true' ionization appears.

Acknowledgement

I would like to thank P. Hvelplund and H. Knudsen, with whom I have had a close cooperation.

References

1. H. Damsgaard, H.K. Haugen, P. Hvelplund, and H.Knudsen, Phys.Rev. A 27(1983)112
2. C.L. Cocke, R. DuBois, T.J. Gray, E.Justinoano, and C.Can, Phys. Rev.Lett. 46(1981)1671
3. Tom J. Gray, C.L. Cocke, and E.Justiniano, Phys.Rev.A 22(1980)849
4. A. Müller, W. Groh, and E. Salzborn, Phys.Rev.Lett. 51(1983)107
5. L.H. Andersen, M. Frost, P. Hvelplund, H. Knudsen, and S. Datz, Phys.Rev.Lett. 52(1984)518
6. R. Shakeshaft, Phys.Rev.A 18(1978)1930

7. W. Steckelmacher and M.W. Lukas, J.Phys.E 12(1979)961
8. J.A. Tanis, S.M. Shafroth, J.W. Willis, M. Clark, J. Swenson, E.
   N. Strait, and J.R. Mowat, Phys.Rev.Lett. 47(1981)828; J.A. Tanis, E.M. Bernstein, W.G. Graham, M. Clark, S.M. Shafroth, B.M.
   Johnson, K.W. Jones, and M. Meron, Phys.Rev.Lett. 49(1982)1325
9. D. Brandt, Phys.Rev.A 27(1983)1314
10. M.R.C. McDowell and R.K. Janev to be published
11. F. Folkmann, private communication
12. W. Meckbach, I.B. Nemirovsky, and C.R. Garibotti, Phys.Rev.A 24
    (1981)1793
13. S.D. Berry, I.A. Sellin, K-O. Groeneveld, D. Hoffmann, L.H. An-
    dersen, M. Breinig, S.B. Elston, and M.M. Schauer, IEEE Trans.
    Nucl.Sci. NS30 No 2 (1983) 902
14. R.O. Barrachina and C.R. Garibotti, Phys.Rev.A 28(1983)1821
15. K-O. Groeneveld, W. Meckbach, I.A. Sellin, and J. Burgdörfer,
    Comments At.Mol.Phys. 4(1984)187
16. W. Meckbach, R. Vidal, P. Focke, I.B. Nemirovsky, and E.G.Lepera,
    Phys.Rev.Lett. 52(1984)621
17. J. Macek, , J.E. Potter, M.M. Duncan, M.G. Menendez, M.V. Lucas,
    and W. Steckelmacher, Phys.Rev.Lett. 46(1981)1571
18. R. Shakeshaft and L. Spruch, Phys.Rev.Lett. 41(1978)1037
19. M. Breinig, S.B. Elston, S. Hult, L. Liljeby, C.R. Vane, S.D.
    Berry, G.A. Glass, M. Schauer, and I.A. Sellin, Phys.Rev.A 25
    (1982)3015
20. F. Drepper and J.S. Briggs, H.Phys.B 9(1976)2063
21. C.R. Vane, I.A. Sellin, S.B. Elston, M. Suter, R.S. Thoe, G.D.Al-
    ton, S.D. Berry, and G.A. Glass, Phys.Rev.Lett. 43(1979)1388
22. P. Hvelplund et al., unpublished
23. N. Bohr andd J. Lindhard, K.Dan.Vidensk.Selsk.Mat.Fys.Medd. 28
    (1954) No 7
24. H. Knudsen, H.K. Haugen, and P. Hvelplund, Phys.Rev.A 23(1981)579

L-SHELL VACANCY PRODUCTION BY ELECTRON CAPTURE TO PROJECTILE-CENTERED CONTINUUM STATES (ECC) IN PROTON-ARGON COLLISIONS

L. Sarkadi*, J. Bossler, R. Hippler, H.O. Lutz

Fakultät für Physik, Universität Bielefeld, Fed. Rep. Germany

*Institute of Nuclear Research of the Hungarian Academy

of Sciences (ATOMKI), H-4001 Debrecen, Hungary

The phenomenon of the electron capture into the continuum states of the projectile (ECC) in ion-atom collisions is widely studied both experimentally and theoretically (see, e.g. Breinig et al. 1982). However, all the studies pertain so far to weakly bound electrons only, and the role of this process in inner-shell vacancy production has not been clarified yet. In previous measurements of L-shell doubly differential (δ-electron ejection) cross sections of argon by proton impact (Sarkadi et al.) we have seen evidence of the existence of the ECC process in inner-shell ionization. To directly verify this process, we have now extended the angular range of our measurements for detection of ejected inner-shell electrons into the forward direction.

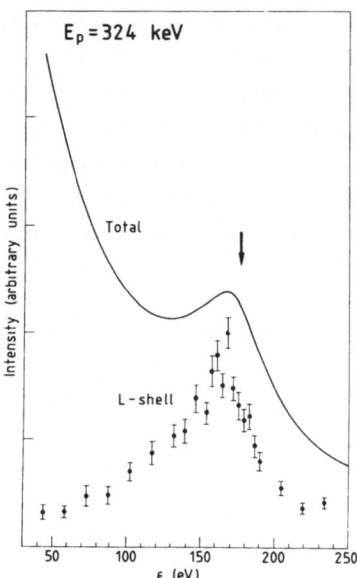

Fig. 1: Energy spectrum of electrons ejected at 0° in 324 keV proton-argon collisions. Singles (total) and coincidence (L-shell) spectra are shown.

We report here on the observation of ECC processes following L-shell ionization of argon by proton impact. This has been achieved by measuring the cusp-shaped peak in the energy spectrum of ejected electrons at 0° (with respect to the incident proton direction) in coincidence with Ar-LMM Auger electrons ejected at 90°. A 45° parallel-plate electrostatic analyzer was used for the detection of δ-electrons. This analyzer was operated with a modest energy resolution of 11%; its angular acceptance was 5.5°. Fig. 1 shows two electron spectra obtained for 324 keV proton impact on argon. The singles (non-coincident) spectrum represents contributions from all shells (predominantly from the M-shell), the coincidence spectrum gives the L-shell contribution only. The pronounced peak in the coincidence spectrum appearing at the same electron energy as the less pronounced peak in the singles spectrum proves the existence of ECC processes from the L-shell. The expected position of the ECC peak as calculated from the projectile velocity is indicated by an arrow. Measurements at an incident proton energy of 278 keV show the expected shift of the ECC peak when compared to the 324 keV measurement. From the measured coincidence rates we have determined absolute doubly differential cross sections (DDCS) for electron ejection at 0°. These DDCS have been compared in Fig. 2 with plane wave Born (PWBA) calculations of Madison and Manson (1979), where the ECC process has been included with the help of Salin's (1972) theory. It appears that Madison and Manson's calculations which exclude the ECC process underestimate the background under the ECC peak by about a factor of 4. The full curve in Fig. 2 is based on Salin's theory and includes the ECC process. To facilitate comparison with the experimental data the calculations have been folded with the experimental angular resolution. Although this results in a cusp-shaped structure in the ejected electron spectrum, in qualitative agreement with the experimental data, we find large discrepancies between experiment and theory which make more refined calculations necessary.

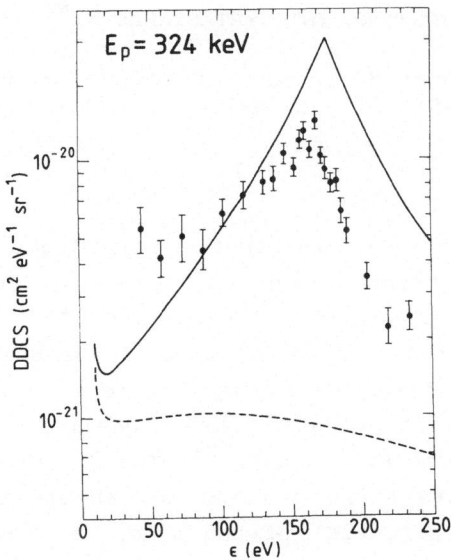

Fig. 2: Doubly differential cross sections (DDCS) versus electron
energy for electron ejection at 0° in 324 keV proton-argon
collisions. The experimental DDCS for the L-shell are
compared with PWBA calculations without (dashed line) and
with (solid line) inclusion of the ECC process (see text).

This work was supported by the Deutsche Forschungsgemeinschaft (DFG).

References

M. Breinig et al. 1982, Phys. Rev. A 25, 3015

D.H. Madison, S.T. Manson 1979, Phys. Rev. A 20, 825

A. Salin 1972, J. Phys. B 5, 979

L. Sarkadi, J. Bossler, R. Hippler, H.O. Lutz 1983. J. Phys. B16, 71

ELECTRON CAPTURE INTO METASTABLE Kr^{8+} RECOIL IONS

Kurt M. Cramon and Finn Folkmann
Institute of Physics, University of Aarhus

Abstract

Energy and time distributions of Auger electrons from highly ionized krypton have been measured after impact on a krypton gas of a pulsed 0.7-MeV/amu Cl^{11+} beam. The energy spectrum, measured delayed relative to the beam, exhibits MNn Auger electron lines following electron capture into metastable $3d^{-1}4s$ excitations of Kr^{8+} recoil ions, with smaller contributions from Kr^{9+} and Kr^{10+}. The distribution of primary-capture orbits in the n=6 and n=7 shells is changed with the ionization potential of the donor gas, observed in gas-mixture experiments. Theoretical calculations of lifetimes and Auger electron energies are made. The classical model for electron capture explains both the mean-excitation state and the qualitative variation of the found total capture cross sections.

1. Introduction

Electron ejection after impact of heavy ions is not always governed alone by the immediate emission from the collision system. Subsequent electron capture into metastable, slowly recoiling ions can add an Auger line spectrum[1]. Time-differential measurements will enable isolation of the delayed capture lines, as shown for Ne^{8+} [2] and for Ar^{8+} and S^{6+} [3], where also measurements of the decay curves for gas mixtures allow determination of the total capture cross sections. This investigation extends the use of the technique of delayed-electron measurement[3] to a case with a large core, a krypton ion, which causes spreading of the various possible capture states (n,ℓ) for different ℓ values of same n shell to be larger than the distance between subsequent n shells. A comparison between experimental and theoretical Auger lines shows that the dominant state, to which the capture takes place, is the metastable $3d^{-1}4s$ excitation of the closed 28-electron state KLM of Kr^{8+}.

2. Experiments

A krypton gas target was bombarded with heavy ions, preferentially 0.7-MeV/amu $^{84}Kr^{4+}$, from the 6-MV EN Tandem van de Graaff accelerator at Aarhus, and electrons were measured perpendicular to the beam
with an electrostatic retarding-field analyzer run with energy resolution 0.9 eV FWHM[2,3]. We used a bunched beam with 5-ns wide beam pa-
ckets separated 500 ns, and measured the time difference between the
channeltron electron signal and the beam-control signal, with a time
resolution of 9 ns FWHM. In the differentially pumped gas cell with
2.0mm×2.0mm entrance and 2.4mm×2.4mm exit holes, the pressure was measured with a membrane gauge to typically 10 to 20 mtorr and the inlet
of one or two gases regulated with control valves and flow monitors
to maintain constant pressure and flow ratio. The beam was stripped by
a 50-μg/cm² carbon foil, 10 mm in front of the gas cell, to a mean
charge state, which for 0.7-MeV/amu Cl was 11+ [4]. The time pulse-
height signal and the electron multiscaling-energy information (in
steps of 0.15 eV) were accumulated in list mode on magnetic tape, and
windows were later set on either time or energy to generate sorted
spectra of the alternative kind.

In Fig. 1 are shown the energy spectra for a pure krypton gas,
both total at the top without conditions on the time measurements, and
prompt just below it at the same scale, with a 140-ns wide window
around the prompt peak. The delayed spectrum from 0.7-MeV/amu Cl^{11+}
impact on 15-mtorr Kr is sorted with a 35-370 ns delayed time window
and represents the difference betweeen the total and the prompt spectrum. It shows clearly the lines of the total spectrum, which are nearly gone in the prompt part. Delayed krypton spectra obtained in the
same way for impact with ions of different type, charge, and velocity
exhibit an identical form, as demonstrated in Fig. 1 for 0.47-MeV/amu
Br^{16+} and 1.0-MeV/amu Cl^{12+}. The bottom spectrum is measured with a
better instrumental resolution of 0.3 eV, but this makes no great improvement in the measurements as the peaks have an internal width exceeding 0.8 eV. The dominant part of the spectrum is a group of at
least four lines between 57 and 62 eV, where the central broad peak
contains at least two peaks.

Time spectra with an energy window around this group of lines are
shown in Fig. 2 for various pressures from 5- to 40-mtorr Kr. These
spectra have an exponential fall-off at long delay times t, being proportional to $\exp(-t/\tau)$, where τ is a characteristic decay time, which
can be fitted reliably at least for pressures below 30 mtorr. From experimental time spectra, τ was fitted in the 125-330 ns delayed region.

From such values, it is possible to obtain information on capture cross sections with some additional assumptions. When a metastable state, with a lifetime long compared to 0.3 µs, captures an electron with a total cross section σ, the capture rate of any of the observed lines

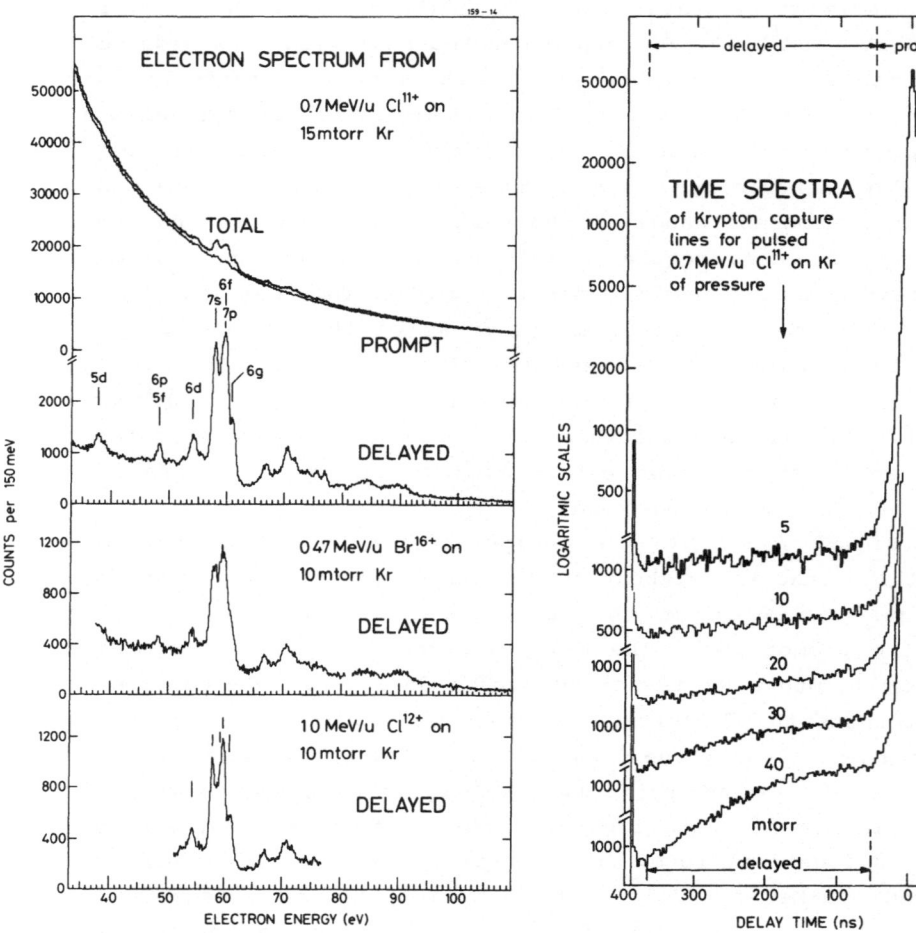

Fig. 1. Electron-energy distribution perpendicular to an ion beam incident on a Kr-gas target. Prompt and delayed spectra are sorted with time windows as in Fig. 2. Bottom spectrum was measured with experimental resolution 0.3 eV FWHM and the others with 0.9 eV FWHM. All delayed lines are from Auger decay of high Rydberg electrons captured into metastable, highly charged Kr ions. nl capture states added to a metastable 28-electron $KLM3d^{-1}4s$ state of Kr^{8+} are indicated.

Fig. 2. Time decay of the group of lines between 57 and 62 eV seen in Fig. 1 for Kr of pressure 5 to 40 mtorr bombarded by a pulsed 0.7 MeV/amu Cl^{11+} beam.

will be $\bar{\tau}^{-1} = \bar{V}_R \cdot n \cdot \sigma$, where $\bar{V}_R$ is the mean velocity of the recoil ion and n the density of neutral atoms, from which the electron can be captured.

With admixture to krypton of other gases, electrons may be captured into highly charged Kr ions, not only from Kr but also from the admixed gas. The delayed electron-energy spectra from Kr mixtures shown in Fig. 3 exhibit such a pattern, where new peaks arise with admixtures of various gases in a systematic dependence on the ionization potential of the added gases.

This selective excitation of various Auger lines in Fig. 3 suggests the presence of a metastable particle-hole state, to which electrons are captured in high Rydberg states, as found for two-electron $K1s^{-1}2s$ ^3S metastables of nitrogen and neon[1,2] and ten-electron metastables $KL2p^{-1}3s$ $^3P_{0,2}$ of argon and sulphur[3]. From the measured capture rates, total cross sections from the admixed gas may easily be found relative to that from krypton alone.

The energies of the Auger lines of Figs. 1 and 3 were found experimentally by comparison with electron-excited Ar LMM Auger electrons and determination of the energy shift with variation of the pressure of the gas, for both 2.5-keV e^- on Ar and 1.0-MeV/amu Cl^{12+} on Kr. As indicated in Fig. 4, the position of the main-peak changes with the charge cloud in the collision region, i.e., as function of pressure times current. This relation is linear for electron impact of argon, determining the energy E_0 with zero retarding voltage. For ion impact on argon, the relation was also linear, but for ion impact on krypton, there was found no obvious linearity at low pressure, but emphasizing measurements in the 10-20-mtorr range, we have drawn a line giving $E-E_0 = 29.5$ eV at zero pressure for the main peak. The other spectra are matched to absolute scale by an easily distinguishable, narrow line such as that at 58.1 eV or 48.3 eV.

3. Theoretical Calculations

To interpret the Auger lines, we shall have to guess both which are the metastable states and calculate the corresponding transition energies.

The best guess on a metastable state is a particle-hole excitation in a filled shell. For krypton, the filled K, L, and M shell contains 28 electrons, and for a $3d^{-1}4s$ excitation of Kr^{8+}, we have calculated the lifetimes from the first-allowed one-photon decay rate, with the multiconfigurational Dirac-Fock program of Grant et al.[5,6] and shown the results in Table 1. We also show the similar calculation for the metastable $2p^{-1}3s$ excitations of the ten-electron KL core of Ar^{8+} and S^{6+}, studied in[3]. The E2-decay branches of the $^3D_1$ state of Kr^{8+} and the $^3P_0$ states of Ar^{8+} and S^{6+} contribute less than 1 per mill of the total decay rates. It is noted from Table 1 that all $3d^{-1}4s$ states in Kr^{8+} have lifetimes exceeding 1.5 μs, and that the ^3D lifetimes alone exceed 4.3 μs, so they are all metastable seen from the point of view of our experiment.

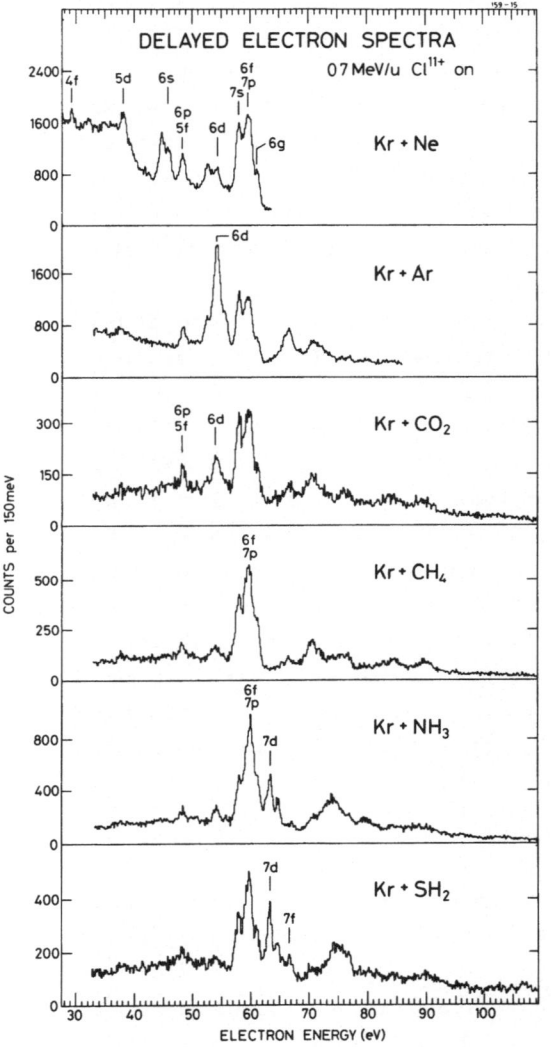

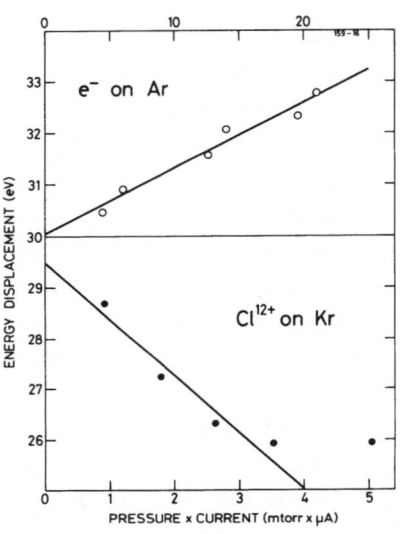

Fig. 4. Energy shift of Auger lines as function of target pressure times current for 5, 10, 15, 20, (25), and 30 mtorr Ar and Kr by impact of 2.5 keV electrons and 1.0 MeV/u Cl^{12+}, respectively. The upper curve gives the zero-pass energy E_0 and the lower curve $E-E_0$ for the main capture line.

Fig. 3. Krypton Auger electron spectra from gas mixtures measured 50-375 ns delayed. Target pressure was 10 mtorr Kr and 10 mtorr of the indicated admixed gas. The selective excitation of special states is seen. The capture orbit added to a $KLM3d^{-1}4s$ state is indicated on the main peaks.

Also other charge states and and electron configurations have metastable states, which may contribute to our spectra, for example has
Kr^{9+} KLM3d^{-2}4s states with E2 lifetimes in the range 276 μs to 0.4
μs, with the main part around one μs, and Kr^{10+} has a broad band of
KLM3d^{-3}4s states, with E2 lifetimes somewhat smaller, but still with
many in the μs range.

Ion	Electron conf.	From	To	Mode	Lifetime	E_M eV
Kr^{8+}	$1s^2 2s^2 2p^6 3s^2 3p^6 3d^9 4s$	$^3D_3$	$^1S_0$	M3	635 s	84.7
	KLM3d^{-1}4s	$^3D_1$	$^3D_2$	M1	93 ms	86.0
		$^3D_2$	$^1S_0$	E2	4.3 μs	85.6
		$^1D_2$	$^1S_0$	E2	1.5 μs	86.0
Ar^{8+}	$1s^2 2s^2 2p^5 3s$	$^3P_2$	$^1S_0$	M2	0.48ms	250.3
	KL2p^{-1}3s	$^3P_0$	$^3P_1$	M1	18.2 ms	252.5
S^{6+}	$1s^2 2s^2 2p^5 3s$	$^3P_2$	$^1S_0$	M2	2.3 ms	169.1
	KL2p^{-1}3s	$^3P_0$	$^3P_1$	M1	135 ms	170.3

*Table 1. Metastable states and their calculated single-photon lifetime and excita-
tion energy E_M relative to the closed shells according to Dirac-Fock programs (Refs. 5-7).*

To find the Auger electron energies, we have used the multiconfigurational Dirac-Fock program of Desclaux[7] as in previous publica-
tions[1,3] and give the excitation energy relative to the KLM ground
state, both of the metastable states in Table 1 and of the capture
states in Table 2, which in addition to the KLM3d^{-1}4s core state have
an n,ℓ electron. The selection rule for the Auger decay in an L-S coup-
ling scheme is that only doublet states with L=ℓ and L=ℓ±2 contribute.

Capture orbit	Auger-decaying multiplet	Core	Auger energy in eV for				
			n=4	n=5	n=6	n=7	n=8
ns	^2D	^3D	-43.0	20.4	45.4	58.4	65.9
		^1D	-41.8	19.4	45.5	58.7	66.3
np	^2P ^2F	^3D	-26.9	26.6	48.3	59.7	66.4
		^1D	-26.3	27.4	49.0	60.7	67.3
nd	^2S ^2D ^2G	^3D	3.3	37.9	54.0	63.0	68.5
		^1D	5.3	38.6	54.9	63.9	69.4
nf	^2P ^2F ^2H	^3D	28.0	48.5	59.5	66.4	70.7
		^1D	29.6	49.2	60.5	67.2	71.6
ng	^2D ^2G ^2I	^3D		49.8	60.5	66.8	71.1
		^1D		50.7	61.3	67.8	71.9

*Table 2. Theoretical Auger electron energies for Kr 29-electron states
KLM3d^{-1}4snℓ decaying to KLM. The Auger decaying multiplets are
indicated, and a mean value is given for both a ^3D and a ^1D state of
the KLM3d^{-1}4s core.*

Therefore we extract from the many calculated states a mean value of such states and pay most attention to those with a ^{3}D core, and the lower transition energies. We also give the energies for similar states with a ^{1}D core, which may also represent a measure for the spreading among the ensemble of capture states for that orbit. It is seen from Table 2 that shells with different n overlap.

The classical model for electron capture outlined in[8] allows determination of the most probable capture orbit and of the capture cross section for a gas of a certain ionization potential I_p. A solid curve showing these values is shown in Fig. 5, employing Eqs. (8) and (9) from[8], using for the ionization energies the ^{3}D values of Table 2 minus the excitation energy of the metastable state (e.g., 86 eV) from Table 1. In Fig. 5 is also shown the unshielded hydrogenic estimate as a dashed curve and the continuum limit as a dotted curve.

The result of the classical model is that a metastable state of charge q and excitation energy E_M captures electrons of ionization potential I_p to states, which Auger decay with an energy centered around

$$E_A = E_M - |I_p|(1 + \frac{q-1}{2\sqrt{q}+1}) , \tag{1}$$

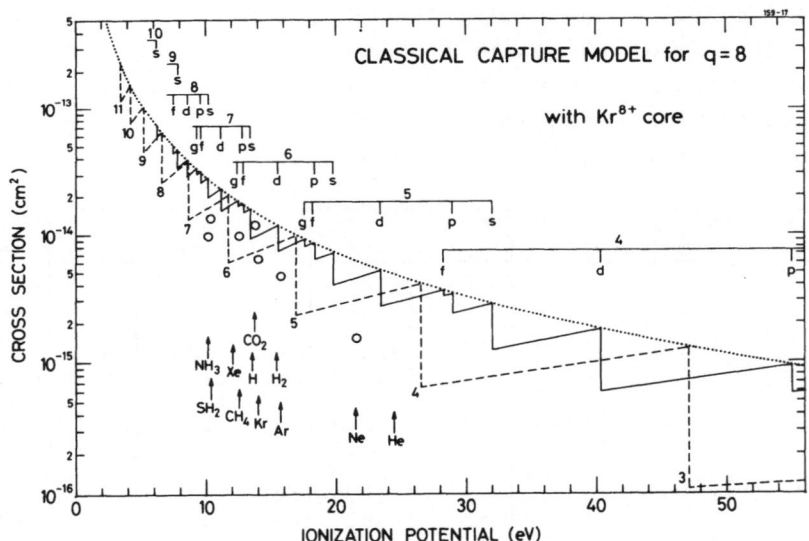

Fig. 5. Classical calculation of capture cross sections and orbits according to Ref. 8, with theoretical binding energies for Kr^{8+} from Table 2 as solid curve. The dashed curve has unscreened hydrogenic energy levels, and the dotted curve indicates the continuous maximum. Ionization potentials of gases of interest are marked with arrows, which point both to the theoretically most probable capture state indicated above the curve and to experimental cross sections from Table 3 (shown as circles).

where the actual orbits are then quantized according to possible values, as for Kr^{8+} given in Table 2.

4. Discussion of the Results

The major part of the measured delayed krypton Auger electron
lines of Figs. 1 and 3 can be understood from the theoretical values
in Table 2, i.e., as resulting from selective electron capture into
the metastable 28-electron state $KLM3d^{-1}4s$ of Kr^{8+}.

The total-capture cross sections are estimated from measurements
of decay times, as shown in Fig. 2. For a mixture of n_1 atoms/cm³ Kr
with capture cross section σ_1 and n_2 molecules/cm³ of an admixed gas,
with capture cross section σ_2, the measured decay rate for capture
staters is

$$\frac{1}{\tau} = \bar{V}_R (n_1 \cdot \sigma_1 + n_2 \cdot \sigma_2) , \qquad (2)$$

where the mean recoil velocity $\bar{V}_R$ depends on the impact parameters
b of the primary collisions[9] by

$$V_R = \frac{Z_t^* Z_p^*}{m_t} \frac{e^2}{b} \sqrt{\frac{2m_p}{E_p}} \qquad (3)$$

for a projectile of mass m_p, energy E_p, and effective charge Z_p^*. For
0.7-MeV/amu Cl, we have $Z_p^* = 11$ and $e^2 = 1.44 \cdot 10^{-10}$ keV·cm. As the
essential criterion for production of the metastable Kr^{8-} states
$KLM3d^{-1}4s$ is that we ionized the 3d shell of neutral Kr in a collision,
we use for Z_t^* the Slater value[10] for the screening of the 3d electrons, $Z_t^* = Z - 21.15$. To estimate b, we use the hydrogenic mean value for the radius of the orbit n,ℓ [10],

$$\bar{r}_n = \frac{1}{2}(3n^2 - \ell(\ell+1))\frac{a_o}{Z_t} , \qquad (4)$$

where $a_o = 5.29 \cdot 10^{-9}$ cm, for the 3d orbit with n,ℓ = 3,2. With these
approximations we obtain $\bar{V}_R = 3 \cdot 10^5$ cm/s or a mean recoil energy $\bar{E}_R$ =
3.9 eV. With this somewhat uncertain value of $\bar{V}_R$, we have dared from
measured values of τ in Table 3 to calculate total-capture cross sections stated as the last entry. They are reasonably constant for the
various pressures of pure Kr, and using an appropriate mean value,
$0.64 \cdot 10^{-14}$ cm², we also calculate the capture cross sections for admixed gases, which are only determined reliably relative to this value for Kr and with considerable experimental uncertainties. We have
plotted the cross-section values of Table 3 in Fig. 5 as circles according to their ionization potential, and although they are lower
than the classical estimate by a factor of about two, they reproduce
the dependence with I_p of the classical model.

Gas	Pressure mtorr	Decay time τ ns	σ 10^{-14} cm²
Kr	5	2357±752	0.85
Kr	10	1416±180	0.71
Kr	15	1069± 95	0.63
Kr	20	787± 46	0.64
Kr	30	550± 21	0.61
Kr	40	261± 5	0.96
Kr+Ne	10+10	1272±173	0.15
Kr+Ar	10+10	905±107	0.47
Kr+CO₂	10+10	550± 74	1.19
Kr+CH₄	10+10	620± 76	0.98
Kr+NH₃	10+10	628± 64	0.96
Kr+SH₂	10+10	509± 56	1.34

Table 3. Experimental decay times fitted in a 125-330-ns delayed region, and estimates of total-capture cross sections using Eq.(2) and $\overline{V}_R = 3.0 \cdot 10^5$ cm/s. For the gas mixtures, we use a krypton cross section $\sigma_1 = 0.64 \cdot 10^{-14}$ cm² and $\sigma_2 + \sigma_1 = 1.01 \cdot 10^{-11}$ ns·cm²/τ.

Gas (I_p,eV)	Energy (eV)	Rel. int.	Identification	Gas (I_p,eV)	Energy (eV)	Rel. int.	Identification
Kr	37.8	0.2	5d	Ar	48.3	0.1	5f,5p
	48.3	0.1	5f,6p		52.7	0.2	
(14.00)	54.3	0.2	6d	(15.75)	54.2	1.0	6d
	58.1	0.6	7s		55.6	0.2	
	59.2	0.4	6f		58.1	0.1	7s
	59.9	1.0	7p		66.7	0.3	7s (9+)
	61.2	0.3	6g,h				
	66.8	0.2	7s (9+)	CO₂	48.3	0.2	5f,6p
	70.6	0.3	6g,7p (9+)		54.2	0.5	6d
	71.8	0.1	6h (9+)	(13.77)	58.1	0.6	7s
	74.1	0.1	7d (9+)		59.8	1.0	7p,6f
	77.0	0.1	7f,8s (9+)		61.2	0.2	6g,h
	84	0.1	7d (10+)				
	89	0.1	7f (10+)	NH₃	58.1	0.2	7s
					59.7	1.0	6f,7p
Ne	29.3	0.2	4f	(10.20)	60.9	0.4	6g,h
	32.1	0.1			63.2	0.6	7d
(21.56)	38.2	1.0	5d		64.6	0.3	8s
	44.9	0.6			73.7	0.5	7d (9+)
	45.8	0.4	6s				
	48.3	0.5	5f,5p	SH₂	59.8	0.6	7p,6f
	52.7	0.3			61.1	0.3	6g,h
	54.2	0.2	6d	(10.40)	63.3	1.0	7d
					64.6	0.4	8s
CH₄	58.1	0.3	7s		66.7	0.3	7f,8p
	59.6	1.0	6f,7p		70.8	0.1	6h,7p (9+)
(12.60)	60.8	0.4	6g,h		75	0.6	7d,f,8s (9+)

Table 4. Measured Kr Auger-line energies and their relative intensity for various studied gases. The ionization potential I_p of the gas is given. The identification gives the capture orbit added to $KLM3d^{-1}4s$ in Kr^{8+}, or when (9+) is given, to $KLM3d^{-2}4s$ in Kr^{9+}, and with (10+), to $KLM3d^{-3}4s$ in Kr^{10+}.

The energies of the krypton Auger-electron lines are found from the delayed spectra of Figs. 1 and 3 and interpreted according to the theoretical calculations in Table 2. The detailed result of this interpretation scheme is given in Table 4, where the electron energies have an uncertainty around 0.2 eV. It is nearly impossible to distinguish between the overlapping 6f and 7p lines. Our experimental energies of the main Kr-capture lines between 57 and 62 eV are in agreement with the observations of Schneider et al.[11] for impact of 1.3-MeV/amu Cl^{12+} and 1.9-MeV/amu O^{5+} on Kr, and in the latter case, Kr-diagram MNN lines in the spectrum enabled an internal energy calibration.

In addition to the strong lines, which are all explained by capture into metastable Kr^{8+}, we observe in Fig. 3 a group of lines between 66 and 82 eV, which exhibit the same kind of variation with the admixed gas as the main lines, and which we interpret by capture into Kr^{9+}. The $KLM3d^{-2}4s$ states of Kr^{9+} are metastable, with an excitation energy around $E_M = 101$ eV above that of the two ground states $KLM3d^{-1}$, and theoretical calculations yield the following Auger energies for capture into 6d, 6f, 6g, 7s, 7p, 7d, 7f, 7g, 8s, 8p, and 8d: 62.7, 68.7, 70.0, 67.6, 69.9, 73.8, 77.5, 78.3, 76.8, 78.2, and 80.6 eV. A higher and even weaker group of lines between 81 and 92 eV, seen best in Fig. 1, can be attributed to capture into metastable Kr^{10+}, for which the $KLM3d^{-3}4s$ states have an excitation energy around $E_M = 118$ eV above that of the nine ground states $KLM3d^{-2}$. Due to the resulting expected spreading of Auger decays to such a band, only broad structures are expected. Theoretical calculations of capture into 7p, 7d, 7f, 7g, 8p, and 8d yield Auger energies of 80.9, 85.0, 89.1, 90.2, 90.7, and 93.4 eV. One confirmation of these observations is that Justiniano et al.[12] explain direct ionization of He by Kr^{8+}, Kr^{9+}, and Kr^{10+} as being due to metastable recoil ions of especially these charge states. Another support to the identifications of the very special excited capture states is that according to Eq. (1), the classical model, e.g., for capture from Kr predicts an Auger energy $E_A = 57.3$ eV to Kr^{8+} with $E_M = 86$ eV, $E_A = 71.0$ eV to Kr^{9+} and $E_A = 86.8$ eV to Kr^{10+}, in close agreement with the observed groups. For capture from Ar, $E_A = 53.7$ eV to Kr^{8+} and $E_A = 67.2$ eV to Kr^{9+}. We have examined the decay curves of all isolated lines in the spectra without finding significant differences from those of the main lines.

The classical model shown in Fig. 5 explains well which states are actually populated by capture of an electron into a metastable Kr^{8+} ion, as seen from the experimental results of Fig. 3 or Table 4.

However, the simple model assumes only a single active state, and a range of states centered around the indicated region is always observed experimentally. As encountered by identification of these lines and seen from Fig. 5, the ℓ splitting within a certain n shell is greater than the distance between adjacent n shells, resulting in ambiguous spectral identification and a spreading of the many possible capture states to approach the continuum maximum indicated in Fig. 5. This is caused by the large 28-electron core of Kr^{8+}. For the ten-electron cores of Ar^{8+} and S^{6+}, the states move halfway between the hydrogenic values[3], and for the two-electron core of Ne^{8+}, the change is only modest[8]. In all cases, the high selectivity of the capture orbit (n,ℓ) in dependence of the ionization potential was demonstrated, and the present investigation exemplifies a situation which approaches a continuum distribution of states but still allows identification of distinct orbitals.

Acknowledgement

This work was supported by the Carlsberg Foundation.

References

(1) R. Mann, F. Folkmann, H.F. Beyer: J.Phys.B 14, 1161 (1981)
(2) F. Folkmann, H.F. Beyer, R. Mann, K-H. Schartner: Nucl.Instrum. Methods 181, 99 (1981)
(3) F. Folkmann, K.M. Cramon, R. Mann, H.F. Beyer: Physica Scripta T3, 166 (1983)
(4) V.S. Nikolaev, I.S. Dmitriev, Phys.Lett. 28A, 277 (1968)
(5) I.P. Grant, B.J. McKenzie, P.H. Norrington, D.F. Mayers, N.C. Pyper: Comput.Phys.Comm. 21, 207 (1980)
(6) I.P. Grant: J.Phys.B 7, 1458 (1974)
(7) J.P. Desclaux: Comput.Phys.Comm. 9, 31 (1975)
(8) F. Folkmann, R. Mann, H.F. Beyer: Physica Scripta T3, 88 (1983)
(9) R. Mann, F. Folkmann, R.S. Peterson, Gy. Szabo, K-O. Groeneveld: J.Phys.B 11, 3045 (1978)
(10) J.C. Slater: Quantum Theory of Atomic Structure (McGraw-Hill, New York, 1960) Vol. 1, p. 369 and 466
(11) D. Schneider, B.M. Johnson, B. Hodge, C.F. Moore: Phys.Lett. 59A, 25 (1976)
(12) E. Justiniano, C.L. Cocke, T.J. Gray, R. Dubois, C. Can, W. Waggoner, R. Schuch, H. Schmidt-Böcking, H. Ingwersen: Phys.Rev.A 29, 1088 (1984)

THREE DIMENSIONAL CONVOY ELECTRON VELOCITY DISTRIBUTIONS PRODUCED BY 60-270 keV PROTON IMPACT ON CARBON FOILS

W. Meckbach, I.B. Nemirovsky and P.R. Focke

Centro Atómico Bariloche - Comisión Nacional de Energía Atómica

8400 - Bariloche, Argentina

The electrons emitted at the exit surface of a foil traversed by an ion beam which have velocities $\vec{v}$ close to the velocity of the emerging ions, $\vec{v}_i$, are called beam foil convoy electrons (BFC). The velocity distributions of these electrons show a cusp-like peak centered around $\vec{v}_i$. Similar cusps are observed in the convoy electrons emitted from gas targets and related to electron loss to the continuum (ELC) and electron capture to the continuum (ECC) mechanisms. In these cases the measured cusp can be discussed in terms of a Coulomb factor in the double differential cross section:

$$\frac{d\sigma}{d\vec{v}} = \frac{1}{v'} \; F(\vec{v},v_i) \;\; ,$$

the primed variable refers to the moving proyectile frame. The origin of BFC elecrons is not well understood though there are many proposed mechanisms (1)(2), and evidence of a $1/v'$ type behaviour (3).

We have measured the BFC electron distributions as a function of v and θ, the emission angle with respect of the beam direction, and compared them with $1/v'$ cusp predictions. In a previous paper we compared such distributions with ELC and ECC ones measured under identical experimental conditions (4). The BFC electron distributions show similarities with a $1/v'$ cusp, though some differences persist. This observation suggests us that the messured BFC electrons may originate close to the surface.

We used a proton beam of energies ranging from 60 keV to 270 keV incident on carbon foils of $2\mu gr/cm^2$. The electron analyzer and the overáll experimental setup used was described previously (4) (5). Our electron analyzer permits measurements at small emission angles $\theta \neq 0°$ by turning the analysing plane through an angle φ. If the instrument is properly adjusted on the beam direction then the relation between θ and φ is:

$$\sin \left(\frac{\theta}{2}\right) = \sin (42.3°) \sin \left(\frac{|\varphi|}{2}\right) .$$

With the foil normal to the beam direction we checked the symmetry of the spectra for $\varphi > 0°$ and $\varphi < 0°$. The spectra were taken with an angular acceptance $\theta_0 = 1°$.

These spectra $Q(E_e, \theta)$ measured as a function of the electron energy E_e and θ are transformed into $Q(\vec{v})$, as a function of the velocity. To account for the distortion introduced by the resolution volume of the instrument (5) we define an experimental double differential secondary electron emission coefficient:

$$\frac{d\gamma_{ex}}{d\vec{v}} \propto \frac{Q}{v^3} .$$

If $d\gamma/d\vec{v}$ is the real double differential electron emission, then:

$$\frac{d\gamma_{ex}}{d\vec{v}} \simeq \frac{d\gamma}{d\vec{v}}$$

when $d\gamma/d\vec{v}$ does not change significantly within the resolution volume in $\vec{v}$ space. This identity is not conserved near the BFC electron peak where $d\gamma/d\vec{v}$ is rapidly varying.

The "transverse" distributions $\frac{d\gamma_{ex}}{d\vec{v}}$, for $v = v_p$, where $v_p \simeq v_i$ is the velocity at the peak measured at $\theta = 0°$, show a rounded top in disagreement with the well known sharp cusp seen in the "longitudinal" distributions $\frac{d\gamma_{ex}}{d\vec{v}}$ with $\vec{v} \| \vec{v}_i$. The FWHM of the transverse distributions ΔV_t is proportional to v_p and equal to $0.06\ v_p$. We also determined the width ΔV_t as a function of θ_0 for protons of 180 keV, and θ_0 ranging from $0.5°$ to $2.5°$, the width increases linearly with θ_0, but ΔV_t remains finite when $\theta_0 \rightarrow 0$. For a $1/v'$ cusp the behavior is $\Delta V_t = 2.32\theta_0$. v_p in agreement with ELC and ECC spectra obtained under similar experimental conditions using a He gas target (6). In these considerations we have not substracted any background (3)(5).

Using spectra measured at different angles θ we constructed three-dimensional representations in $\vec{v}'$ space and contour lines taken at different levels as shown in Fig. 1 a,b. We see that for levels close to the peak these lines suffer a transverse elongation that reflects

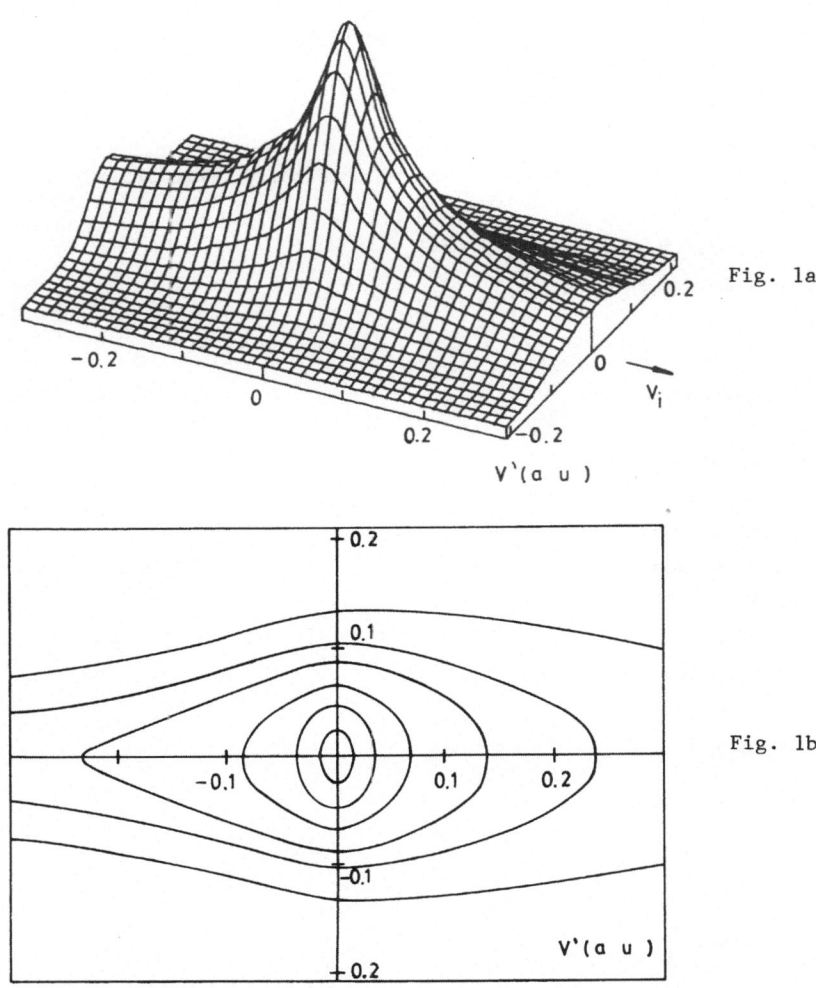

Fig. 1a

Fig. 1b

Fig. 1a) Three dimensional measured BFC electron cusp obtained with 98
keV protons interacting with a carbon foil, represented in
$\vec{v}' = \vec{v} - \vec{v}_p$ space. Peak velocity $v_p = 1.98$ au. Acceptance angle
$\theta_0 = 1°$.
  1b) Contour lines obtained from the cusp taken from inside out at
0.9, 0.7, 0.5, 0.3, 0.2 and 0.1 fractional levels of the peak
height.

ANOMALOUS MEAN FREE PATHS FOR SCATTERING OF CONVOY ELECTRONS GENERATED BY FAST, HIGHLY IONIZED IONS IN THIN SOLID TARGETS*

I. A. Sellin, S. D. Berry, and M. Breinig

Dept. of Physics, University of Tennessee, Knoxville, TN 37916

and

Physics Division, Oak Ridge National Laboratory, Oak Ridge, TN 37831

C. Bottcher

Physics Division, Oak Ridge National Laboratory, Oak Ridge, TN 37831

R. Latz, M. Burkhard, H. Folgert, H. J. Frischkorn, K. O. Groeneveld,

D. Hofmann, and P. Koschar

Institut für Kernphysik der Universität Frankfurt/M

Federal Republic of Germany

Recent work by Betz et al. [1] and by Yamazaki and Oda [2] has emphasized the role of the potential step and sudden shielding change on high Rydberg state production and convoy electron production for fast, heavy particles exiting solid targets. An issue in question is whether the final states are formed at the surface on exit, or in the bulk over some characteristic distance before exit. All recent studies incorporate the latter suggestion, which is at first sight surprising, since it involves long term association of the electrons and the projectile over a long path length.

Several puzzling aspects of convoy electron production by fast projectiles traversing foil targets have been noticed. Chief among these are difficulty in reconciling the strong Z-dependence we observed in our laboratory for the total convoy electron yield with coincidence yield data for emergent $C^{(4-6)+}$ and $O^{(5-7)+}$ ions traversing various solid targets, which showed the observed yields to be nearly independent of emergent ion charge q_e. In channeling experiments [3] the data indicated a precursor electron capture inside the solid, probably to an excited n-state, quickly followed by loss of such electrons to convoy states.

What was difficult to understand about these channeling coincidence experiments as well as similar experiments with amorphous solid targets was the mechanism which reduced the correlation between the convoy electron yield and q_e. Mean free paths for free electrons of the velocity in question were known to be ~20Å, whereas the mean free path for projectile charge change was ~200Å. Thus, an ion was unlikely to change charge in the 20Å layer from whence the observed convoy electrons were thought to come (most others scatter out of the cone of observation).

Additional puzzles are posed by data taken using light projectiles. For hydrogen and helium projectiles, the yields of convoy electrons mimic the background electron yield [2,4], but do not track projectile charge state evolution. Other investigators [5,6] measuring the energy distribution of convoy electrons using mostly hydrogen and helium projectiles note two components in the observed electron velocity distribution (a sharp feature atop an underlying helmet-shaped hump, offset from the center of the hump). This two-component nature of the cusp in ejected electron energy (longitudinal velocity) has not been observed in work with heavy projectiles. A second cusp component of different origin [4], arising from field ionization in the electrostatic spectrometer, normally used by us and others, has been shown to be an artifact of the particular apparatus used in some studies [7,4].

Meanwhile, the production of Rydberg states of projectiles emerging from solid targets has been investigated by several authors. For heavy projectiles, it is found [1] that Rydberg states of high ℓ are plentiful, contrary to the predictions of a single atom pickup model (e.g. calculated in the first Born approximation). However, no significant initial nonstatistical sublevel population has yet been observed [7], in contrast to proton data which apparently show strong indication of alignment in the excitation [4].

A partial but incomplete resolution of the puzzles concerning the n and ℓ-state distribution of foil-excited Rydberg states was achieved in work by Betz et al. [1] and by Yamazaki and Oda [2]. These authors have emphasized the possible importance of the sudden change in projectile shielding as the projectile exits; the ion sees a potential step ~ $Ze^2\omega_p/v$, where ω_p is the bulk plasmon frequency of the target, and Z and v describe the projectile atomic number and velocity, respectively. Betz et al. trace the origin of such high Rydberg-state production to the electrons transferred from the continuum to high n-and ℓ-states at this potential step; Yamazaki and Oda invoke the same mechanism to explain the generation of what they term intrinsic convoy electron production (the spike on the helmet) in terms of free electron transfer to the sharp feature from the underlying secondary electrons.

In an attempt to sort out these problems, we have measured the yields of convoy electrons associated with 15.2 MeV/amu Ni^{24+} and Ni^{26+} ions transversing C and Al. By observing how the yield saturates with increasing thickness, we obtained effective mean free paths (MFPs) for the scattering of convoy electrons out of the collector half angle (1.54°). Figure 1 shows a typical plot. The MFPs thus derived are about 10 times those expected for free electrons of the same velocity! - e.g. at an equivalent electron energy of 8.3 keV we find a convoy electrn MFP of ~ 200 ± 20 nm in carbon, as opposed to the free electron value of ~ 10 nm.

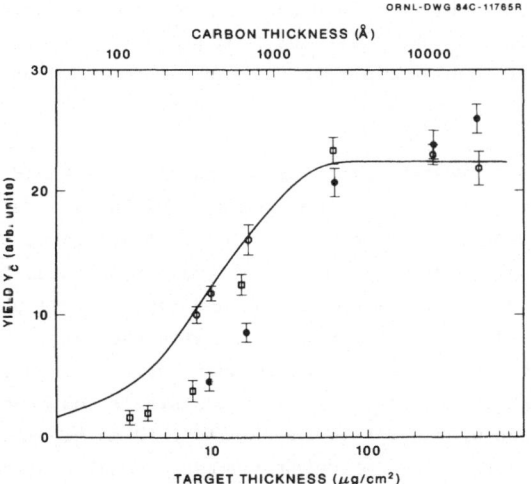

Fig. 1. Convoy electron yield vs. target thickness. The open circles and squares refer to Ni^{24+} ions at 15.2 MeV/u in aluminum and carbon foils and the closed circles to Ni^{26+} in aluminum.

In the present experiments Ni^{24+} and Ni^{26+} ions of 15.2 Mev/u energy were directed through polycrystalline C and Al foils of thicknesses 3-500 μg/cm². The methods used have been described fully by Latz et al. [6] and need not be repeated here. The unique beam used was made available on short notice by the staff of the GSI accelerator at Darmstadt.

The observed convoy electron yields (in arbitrary units) are plotted vs. target thickness in Fig. 1. The absolute yields were consistent with the universal formula of Laubert et al. [8]. The error bars represent counting statistics and uncertainties in extracting underlying electron backgrounds from the observed cusps; the nuclear radiation background contributes a considerable fraction of the uncertainty. Our experiments track the rise in convoy production from effectively very thin targets to as thick as desired. Thus a yield growth curve is measured from which a crude estimate of the MFP can be obtained by fitting to 1-exp(-x/λ), where x is the target thickness, and λ the effective MFP for convoy production. While such a simple estimate is not expected to be accurate, even at the level of a factor of two, the size of effect observed is so large that there can be no doubt as to its significance. The approximate independence of the earlier coincidence data now becomes easier to understand, since the yield could be influenced by the charge state history of the projectile quite deep in the foil.

We have examined the interesting suggestion [9] that the effect might be due to parasitic dependence on electron loss from the projectile as it penetrates the foil, as the L- and M-shell electrons approach charge state equilibrium in the foil thickness and projectile velocity range in question. However, the electron loss cross section from Ni^{24+} (2 L-shell electrons) is almost an order of magnitude larger [10] than for Ni^{26+} (only K-shell electrons) while our measured yields are comparable.

Correlations between the motions of the projectile and electron over distances of several thousand Å may seem surprising but less so when the magnitude of the electric field near a 26+ ion is considered, together with the long plasma shielding distance (~ 100 a_0) at these velocities [11]. The electron-projectile distances at which loss and capture take place are also ~ 10-100 a_0. We visualize an electron moving with the velocity originally equal to that of the ion beam p_0. (Atomic units will be used so that m = 1.) Collisions with target atoms create a small transverse component $\vec{p}_\perp$ which does a random walk combined with drift in the ionic field as electron and projectile proceed through the foil. The following heuristic considerations indicate that such a model is quantitatively reasonable.

Electrons are detected when their net scattering angle is within the detector half-angle θ_D. If the average scattering angle at each collision is θ_s, and the total number of scatterings (about one per atomic layer of the foil) is N_s, the net angle $N_s^{1/2}\theta_s$, whence θ_s ~ 0.05-0.02 deg for 10^3-10^4 scattering layers. The observed $\theta_s = \theta_s^0 - \theta_c$, where θ_s is the deflection in one target collision, and θ_c is the "Coulomb focusing angle" through which the electron is pulled back between scattering events by the electric field F of the projectile. If the interlayer distance is a and the projecile velocity is v, the electron changes its transverse velocity by Fa/v between layers, so that θ_c ~ Fa/v^2 [12]. The inverse MFP for scattering through angles > θ_{min} is proportional to $\int_{\theta_{min}}^{\theta_{max}} \sigma(\theta)\sin\theta\,d\theta$ and thus to θ_{min}^{-2} if we assume Rutherford scattering (σ ~ θ^{-4} at small angles) and $\theta_{max} \gg \theta_{min}$. The MFP is thus enhanced by a factor $\zeta = (\theta_s^0/\theta_s)^2 = [1 + (\theta_c/\theta_s)]^2$. The mean force F entering this formula is difficult to specify without detailed calculations, but in the region of charge states q and ionic speeds v such that v < q we can take Bohr's estimate that F $\geq$ binding force of a valence electron (~ 0.5 a.u. in C, Al) for pickup to occur. Thus θ_c ~ 0.1 deg and ζ ~ 9-36 in the present experiments. If our forumula is applied to O^{8+} at ~ 2 MeV/amu (the energies used in Ref. 3) a similar enhancement is found, suggesting a MFP $\geq$ 200 Å, comparable to the MFP for ion charge changing. Thus, the correlation with q_e should be reduced, assisting the interpretation of the concidence data. In the region v $\gg$ q, F is much smaller and ζ ~ 1, in harmony with Ref. (4) which finds a normal MFP for protons at v = 5 a.u.

A rigorous formulation of the model may be based on a transport equation for the time-dependent distribution of the electrons in position and momentum space $\psi(\vec{r},\vec{p};t)$. If we introduce a phenomenological diffusion coefficient D the time evolution of ψ is described by the transport equation,

$$\frac{\partial \psi}{\partial t} = -\vec{p} \cdot \frac{\partial \psi}{\partial \vec{r}} - \vec{F}(\vec{r}) \cdot \frac{\partial \psi}{\partial \vec{p}} + D \frac{\partial^2 \psi}{\partial \vec{p}_\perp^2} . \tag{1}$$

Assuming a central force and cylindrical symmetry, Eq. (1) is a parabolic partial differential equation in 5 dimensions other than time. A reasonable approximation might be to retain only the radial cylindrical components of $\vec{p}$ and $\vec{r}$, but the solution of even this equation is a large computational project. A more drastic approximation is to drop the first term in the RHS of Eq. (1), i.e. to neglect transport in space, an assumption probably valid when F is not too large, say for $r >$ some r_0. This does lead to an analytical solution which has a physically appealing interpretation. The momentum distribution at first spreads out into a Gaussian of width $\sim (Dt)^{1/2}$ while the number of electrons decreases exponentially with time, correponding to a flux through the surface $p = 0$, $r = r_0$ in phase space. It is tempting to identify the flux loss with the formation of Rydberg or wake-riding states; more detailed analysis shows that this loss dominates where $\theta_s < \theta_c$ in accord with the earlier discussion. These states are not, of course, static within the foil, but are continually destroyed and reformed, maintaining dynamic equilibrium; the idea of continuously created and destroyed, but on average populated, quasicontinuum states is familiar in the study of dense plasmas [13].

Our final picture is thus of a net flux into quasi-bound states due to the attractive field of the ion; on exit these states have a moderate-to-high probability of ionizing in the potential step and producing a cusp electron yield corresponding to long MFP's. We thus succeed in unifying and extending the ideas of Betz [1] and Yamazaka and Oda [2] concerning the association of electrons and projectile within the foil and the role of the potential step on exit. Our next task is the numerical solution of Eq. (1) with some allowance for plasma shielding and the wake potential to check the simple arguments advanced above and to guide the selection of future experiments.

*The measurements were made at GSI UNILAC. Research of CB was sponsored by the U.S. Department of Energy under contract no. DE-AC05-84OR21400 with Martin Marietta Energy Systems, Inc; research for RL, MB, HF, HJF, KOG, DH, and RK was supported by BMFT-Bonn and DFG-Bonn; and research for SE, MB and IAS was supported by the U.S. National Science Foundation, Physics Division and Division of International Programs.

†Permanent address: Gessellschaft für Schwerionenforschung, GmbH, Darmstadt, FRG

1. Betz, H.-D., _et al._, Phys. Rev. Lett. 50, 34 (1983).

2. Yamazaki, Y. and Oda, N., Phys. Rev. Lett. 52, 29 (1984).

3. Breinig, M., Elston, S.B., Huldt, S., Liljeby, L., Vane, C.R., Berry, S.D., Glass, G.A, Schauer, M., Sellin, I.A., Alton, G.D., Datz, S, Overbury, R., Laubert, and Suter, M., Phys. Rev. A25, 3015 (1982); and references therein.

4. Kanter, E.P., Schneider, D., Vager, Z., Gemmell, D.S., Zabranski, B.J., Yuan-zhuang, Gu, Arcuni, P., Koch, P.M., Mariani, D.R. and Van de Water, W. Phys. Rev. A 29, 583 (1984); Vager, Z. et al., Phys. Rev. Lett. 48, 592 (1982).

5. Gladieux, A. and Chateau-Thierry, A., Phys. Rev. Lett. 47, 786 (1981).

6. R. Latz, R. Astner, G., Frischkorn, H.J., Koschar, P., Pfenning, J. Schrader, J. and Groeneveld, K.O., Nucl. Instr. Meth. 194, 315 (1982).

7. Engar, P, Breinig, M., DeSerio, R., Sellin, I.A., Gonzalez-Lepera, C.E. Berry, S. D., Bloemer, M. and Underwood, T., Conference Abstracts, XIII International Conference on Atomic Physics, #41, Seattle, Washington, (July, 1984), and references therein.

8. Laubert, R., Sellin, I.A., Vane, C.R., Suter, M., Elston, S.B., Alton, G.D. and Thoe, R.S., Nucl. Instr. Meth. 170, 577 (1980).

9. Betz, H.-D., private communication.

10. Gould, H., Greiner, D., Lindstrom, P., Symons, T.J.M. and Crawford, H., Phys. Rev. Lett. 52, 180 (1984). Also, private communication with Gould, H.

11. Neufeld, J. and Ritchie, R.H., Phys. Rev. 98, 1632 (1955).

12. Bottcher, C., J. Phys. B 11, 3887 (1978).

13. Flannery, M.R., Case Studies in Atomic Collision Physics 2 , 1 (1972).

H.-D.Betz

Sektion Physik, Universität München, D-8046 Garching

RYDBERG-STATE PRODUCTION IN COLLISIONS BETWEEN FAST

IONS AND CARBON TARGETS

CONTENTS

I. Introduction

It is known for some time that prompt x-rays can be observed from projectile ions long after these ions have been excited by beam-foil interaction[1-7]. Apparently, one detects the decay of long-lived states which
proceeds either directly or via cascading into the K-shell. Since di-
pole transition rates scale approximately $\propto n^{-3} \ell^{-2}$, where n and ℓ de-
note principal and angular momentum quantum numbers of initially exci-
ted states, respectively, it is obvious that for sufficiently high values of n and/or ℓ very long lifetimes must result. A further signature
of these Rydberg state decays becomes evident in the decay curve: the
observed intensity does not vary exponentially with time but exhibits
a power-law dependence[1], $I(t) \propto t^{-a}$, an effect of the addition of contributions from various initial states with different lifetimes and populations.

The question arises how these Rydberg states are formed. In case of a
single ion-atom collision either bound-bound excitation of a projectile
electron or capture of a target electron into projectile states are the
only two relevant processes. In the first case, first order Born approximation calculations are expected to describe the various excitation
probabilities $\sigma(n_i \ell_i \rightarrow n_f \ell_f)$ with reasonable accuracy. In the second
case, precise theoretical predictions are not easily at hand although
the problem of electron capture is being tackled at least since 1927.
Capture cross sections $\sigma_c(n)$ have been tested only for modestly high
n by means of field-ionization techniques[8], and the dependence $\sigma(\ell)$
remained practically uninvestigated except for n=2. Although one may
trust present theories to a large extend it has become clear in recent
years that calculation of a reasonably accurate ℓ-dependence requires
more than first-order approaches.

When we turn attention to beam-foil excitation of Rydberg states one
may argue that the underlying mechanism is essentially the same as in
the case of single ion-atom collisions, namely bound-bound excitation
and electron capture. It has often been argued that Rydberg states can
not exist inside a solid; as a consequence, the required collision processes are thought to take place in the "last layer" so that Rydberg
orbitals can be formed asymptotically as the ion emerges from the foil
into the vacuum. Incidentally, one would not insist that such a last-
layer is identical to a mono-layer but may, perhaps, be represented by

several mono-layers. No quantitative model has ever been put forward
to illucidate this question.

In a recent Letter[9] we have shown that no such last-layer hypothesis
is adequate in explaining observed intensities of x-ray decays of Rydberg states. The data demonstrates that a new mechanism is required which
exhibits the unexpected quality to favor population of high-ℓ states in
these very fast collisions. We discuss various attempts to develop such
mechanisms but it is conceded that a final solution is not yet at hand.

It is not very far-reaching to draw close connections between the production of Rydberg- and continuum states[10]; in fact, it is usually
assumed that appropriately normalized state densities are smooth when
crossing the ionization threshold, i.e. the number of continuum- and
Rydberg electrons can be deduced from each other. If such a concept is
valid, it becomes obvious from the above that the formation of convoy
electrons can not yet be completely understood (let us disregard the
distinct process of electron capture to the continuum, ECC), and that
studies of Rydberg states should have implications for the production
of convoy electrons.

II. Cascading of Rydberg States and Decay Curves

For a given initial population of excited states, $N(n,\ell,t)$, the time-
dependent evolution of these populations can be calculated from the
rate equation

$$dN/dt = \sum_{n'\ell'} \left[N(n',\ell',t)R(n',\ell';n,\ell) - N(n,\ell,t)R(n,\ell;n',\ell') \right], \quad (1)$$

where R signifies radiative transition rates between two states $n'\ell'$
and $n\ell$. Due to present experimental restrictions we do not yet include
magnetic substates; this omission is acceptable when neither angular
distributions of x-rays nor field-quenching effects are considered. The
intensity of any selected transition is obtained from

$$\Delta I = \sum_{n'\ell'} N(n',\ell',t)(1-e^{-R^*\Delta t})R(n,\ell;n',\ell')\Delta t/R^* ,$$
$$\text{where} \quad R^* = \sum_{n'<n} R(n,\ell;n',\ell') . \quad (2)$$

We prefer to study very simple collision systems and choose ion velocities which are sufficiently high to insure that mostly 1-electron
ions are observed. Then, we can utilize well-known prescriptions for
the evaluation of hydrogen-like transition rates. Wit the exception of

the metastable 2s state, which has to be handled separately, we can further limit the treatment to dipole transitions ($\Delta \ell = \pm 1$), whereby the cases $\Delta \ell = + 1$ yield only relatively small contributions. We have developed a numerical code which solves Eq.(1) and (2) for any given initial distribution $N(n, \ell, 0)$ with exact hydrogen-like rates for any range of n- and ℓ-values, provided that n does not exceed ~ 200. The latter restriction is of no practical importance here and is imposed merely by programming techniques and computer capacities. It should be noted, however, that the hypergeometric functions involved in the rates can not be represented by the usual series with alternating terms because calculations would then break down already for $n \cong 20$. Instead, recurrence relations may be exploited to overcome this formal difficulty.

Analytical solutions of the cascade equations (1) and (2) can be worked out for some special cases, in particular for (i) cascading along the yrast line and (ii) summed direct transitions such as ns $\rightarrow$ 2p or np $\rightarrow$ 1s. With the additional assumption of power-low dependences,

$$N(n) = N_0/n^b \quad \text{and} \quad R(n) = R_0/n^c \tag{3}$$

one can easily derive for the two mentioned cases[11]

(i) $\quad \Delta I/\Delta t = R_0 N_0 \left[(1+c) R_0 t \right]^{-(b+c)/(1+c)}$, and $\tag{4}$

(ii) $\quad \Delta I/\Delta t = R_0 N_0 c^{-1} \Gamma (b+c-1)/c \ (R_0 t)^{-(b+c-1)/c}$. $\tag{5}$

As accumulated effect one obtains indeed a power-law dependence for the observed intensity,

$$I(t) \propto t^{-a} \ \text{(Photons/Ion sec)}, \tag{6}$$

where the decay constant a clearly depends on the particular choice of the initial distribution. For b=c=3 one gets a = 3/2 for (i) and a = 5/3 for (ii). A number of sample calculations for varying initial populations have been published[12] and agree with estimates Eq.(4) and Eq.(5).

III. Experimental Aspects, X-Ray Spectra and Decay Curves

The Munich Tandem Van de Graaff facility has been used to accelerate oxygen ions to 60 MeV and sulfur ions to 125 MeV. Sulfur ions were pre-stripped to allow the choice of incident charge states 15^+ and 16^+. Excitation of these beams took place either in carbon foil targets with thicknesses between 2- and 220 $\mu g/cm^2$ or in windowless, differentially pumped gas cells (mostly N_2). X-rays were recorded at 90° with respect to the beam using a flow-mode proportional counter and a Si-(Li) detector for oxygen and sulfur K-shell x-rays, respectively. The distance between target and x-ray observation could be varied in the range $o \lesssim s \lesssim 2m$, corresponding to observation times $t \lesssim 73$ ns. Care was taken to reduce background effects due to undesired excitation of (i) the beam in the residual gas of the beam line and (ii) wall material in the viewing area of the detector resulting from slit-edge scattering of the beam. Absolute x-ray intensities were obtained with the help of well-defined viewing geometry and beam monitoring in a Faraday cup with a biased guard ring to prevent escape of secondary electrons.

Two spectra are displayed in Fig.1 and Fig.2 for sulfur passing through carbon and N_2, respectively, obtained at about 20 cm behind the center of the target. Several K-shell transitions can be identified, as well as a continuous intensity distribution which originates from the 2-photon decay of the metastable 2s-state. The decay length, $s = v\tau(2s) \cong$ 19.8 cm, happens to fall right into the investigated range of distances, but at the position of K-lines the contributions from 2s-decays are small and can be determined quantitatively.

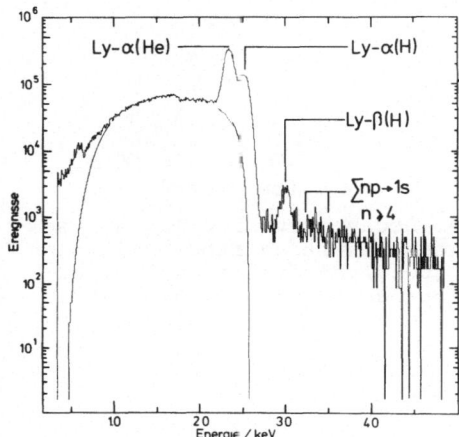

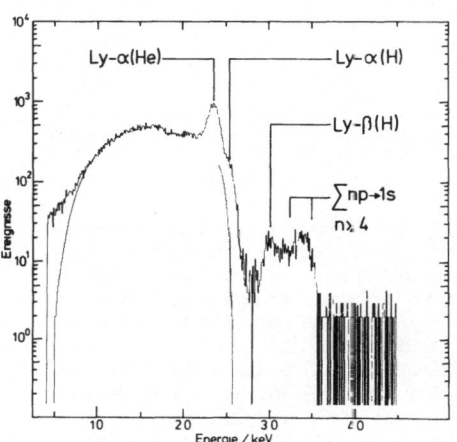

Fig.1: X-ray spectrum for 125-MeV sulfur ions with initial charge state 16$^+$, observed 20 cm behind a 5 μg/cm² carbon target. Three transitions with initial configuration 1s2s+1s2p, 2p and 3p can be identified, as well as the continuous distribution originating from the 2-photon decay of the 2s state.

Fig.2: X-ray spectrum from 125-Mev sulfur ions with initial charge state 16$^+$, observed 20 cm behind a thin N$_2$-gas target. The lines are identified in Fig.1; in addition, relatively strong transitions np→1s with n≥4 become visible.

Decay curves I(t) are shown in Figs. 3-5. For oxygen on carbon we observe a strange form of the decay curve for times $t \lesssim 5$ ns which is difficult to understand on the basis of smoothly populated Rydberg levels[6]; for t > 5 ns, however, the intensity varies as $t^{-1.75}$ without indication of deviations from a power-law dependence.

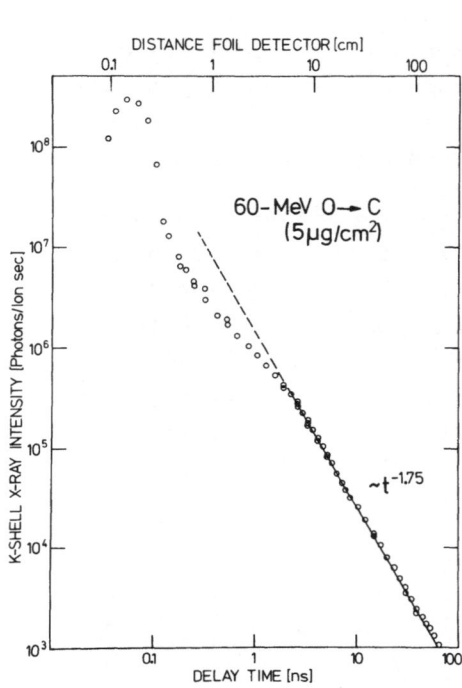

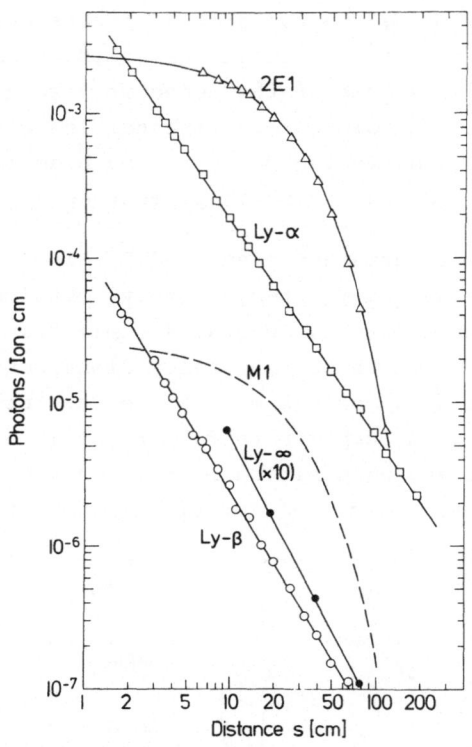

Fig.3: Absolute K-shell x-ray intensity observed for 60-MeV oxygen ions, as a function of time after excitation in a 5 μg/cm² carbon foil, normalized to solid angle 4π. The straight line is a fit to the data for t ⪝ 3ns and represents a power-law dependence with exponent -1.75. For t < 0.1ns geometrical shadowing effects cause a decrease of intensity which is of no physical significance.

Fig.4: Decay curve for 125-MeV sulfur ions resulting from excitation in a 10 μg/cm² carbon foil. Incident charge state was 16⁺.

The two sets of decay curves for sulfur in carbon (Fig.4) and in nitrogen (Fig.5) provide rich information. As expected, the 2 E1 intensity falls off exponentially with the estimated decay length, whereas the Lyman lines exhibit a power-law dependence with exponents varying between 1.5 and 2.0, being constant for a particular transition. In case of gas-excitation it happens that the M1 decay branch of the 2s-state shows an intensity which is comparable with the one of the cascade-fed Ly-α transition. This fact offers the intriguing chance to measure an M1 transition; to our knowledge, such a procedure has not been possible before.

As regards further differences between foil- and gas excitation we point to the strikingly different ratios of Ly-α and Ly-β intensities in these two cases. In the former, we find ratios near 4, whereas in the latter case the ratio varies between 35 and 55 (Fig.6). Inspection of relative strengths of dipole transition rates indicate very clearly the necessary origin of this observed different behaviour and we conclude as follows:

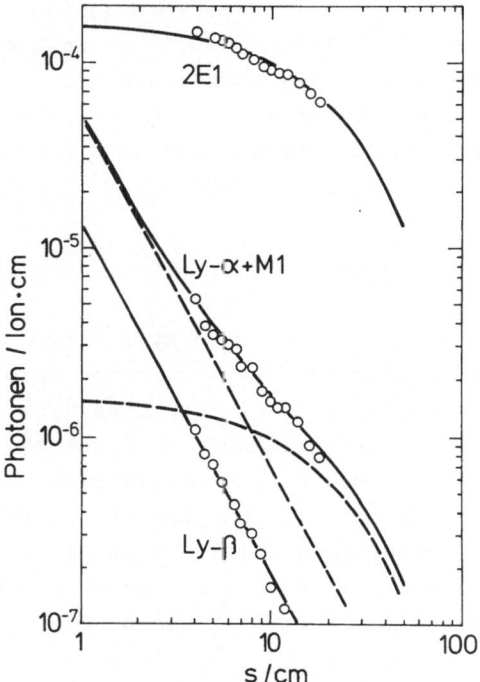

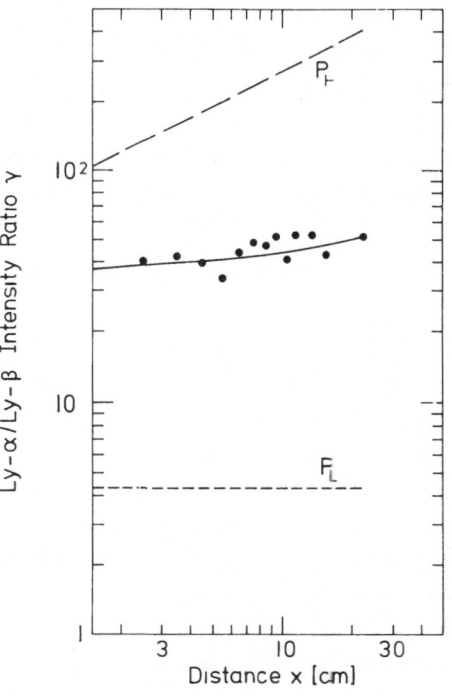

Fig.5: Decay curve for 125-MeV sulfur 16+-ions resulting from electron capture in a thin, windowless gas cell (10^{16} N_2 atoms/cm²). The dashed lines represent a fit of the Ly-α data to a sum of the exponential M1-decay of the 2s state and the power-law like decay of the cascade-fed 2p state.

Fig.6: Ly-α/Ly-β intensity ratio obtained from the data for foil excitation shown in Fig.4 (●). Also given are ratios which should result when only low-ℓ (P_L ---) or high-ℓ (P_H — — —) distributions are assumed as initial Rydberg populations. Experimental ratios from gas excitation (Fig.5) fall close to the line marked P_L.

(i) single ion-atom excitation, whether bound-bound processes for incident 15[+] ions or electron capture for incident 16[+] sulfur ions,
leads to a distribution of Rydberg states which is centered around
low-ℓ values.

(ii) ion-foil induced Rydberg state populations accentuate high-ℓ
states ($\ell \gtrsim 10$).

These situations are also illustrated in Fig.6. We find that (i) is in
reasonable agreement with theoretical calculations whereby our refined
data analysis reveals some sensitivity to the chosen model: there is
evidence that the ℓ-distributions form first order Born approximation
for electron capture is in less satisfactory agreement with our data
compared with the ℓ-distribution from high-order theories such as the
continuum distorted wave method (CDW) as formulated by Dubé[13] or Belkic
et al.[14]. Detailed evaluations of the measured decay curves are still
in progress and will be published separately. Production of high-ℓ states (ii) in ion-foil interaction presents still a puzzle and will be
discussed in the following section.

IV. Target-Thickness Dependence and Capture Model

The "last-layer" hypothesis adressed above has been tested by a very
simple technique[9]: sulfur ions with charge state 16[+] were directed onto
very thin carbon foil targets and the intensity I(t) of cascading Rydberg levels was measured as a function of target thickness,x. Since the
equilibrium charge state is near 14[+] (for 125-MeV S) and equilibrium
thickness is near some 200 $\mu g/cm^2$ one can be certain that the 16[+] charge state fraction, Y^{16+}, decreases for increasing x (Fig.7, top). As a
consequence, when Rydberg states were formed via capture of target electrons during emergence of the ions from the foil, I(t) should be proportional to Y^{16+} and, thus, decrease for increasing x, at least in
a range of small x-values where sort of single-collision conditions pre-
vail. Fig.7 demonstrates, however, that the opposite is true. Together
with the information acquired from experiments with incident 15[+] ions
we summarize a consistent explanation as follows (see Fig.8):

Rydberg states are not primarily formed by direct capture of target
electrons. Instead, Rydberg state production is proportional to the
number of projectile electrons in higher core states ($2 \lesssim n \lesssim 6$) which
are collected either by electron capture or bound-bound excitation in
collisions inside the foil. In a consequent step these electrons be-

come ionized which means that a large fraction of these electrons then travels approximately with projectile velocity along with the ion. During and upon emergence from the foil these electrons end up either as free convoy electrons or as (bound) Rydberg electrons (Fig.8).

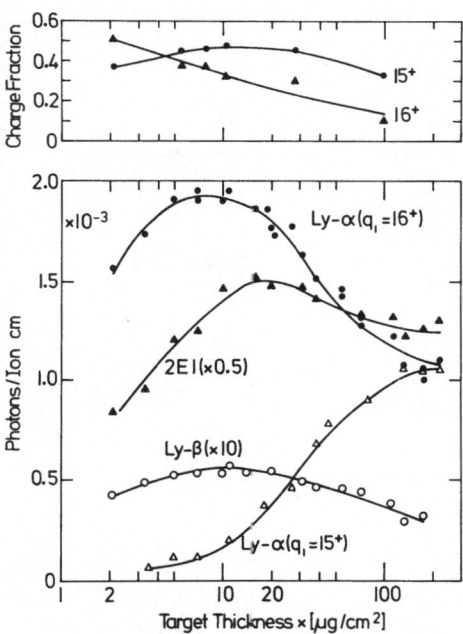

Fig. 7: Absolute K-shell x-ray transition intensities (bottom) and charge-state fractions (top) for 125-MeV sulfur ions, as a function of the thickness of the traversed carbon foil. Incident ions are fully stripped (16^+); Ly-α yields are also shown for incident charge state 15^+. X-ray observation took place 2 cm behind the foils, a distance which corresponds to 1.2×10^5 decay lengths of the prompt 2p-1s transition.

Unfortunately, no precise interaction mechanism can be given which performs the last step of re-capture to the ion with the condition that high-ℓ states are more heavily populated than low-ℓ states. We repeat that neither direct Coulomb capture nor bound-bound excitation can explain the abundance of high-ℓ states in fast collisions, even when additional action of Stark-mixing is admitted.

The produced n-dependence may, perhaps, be close to $N(n) \propto n^{-3}$; we add, though, some words of caution. First, the decay curve for oxygen[6] (Fig.3) rules out such a dependence for quite a range of relatively small n-values. Second, we can present a mechanism which yields a dependence $N(n) \propto n^{-1}$ for a very large range of n-values and offers an ℓ-distribution as is required here. This is radiative electron capture (REC) of emerging convoy electrons by the ions. Fig.9 shows the corresponding cross sections for two relative velocities between electron and ion. The difficulty with this model is the relatively small magnitude of the REC cross section; precise calculations are not yet available and require consideration of the spatial correlation between electrons to be captured and the ions.

In any case, and especially because rough intensity estimates suggest approximately equal probabilities for production of Rydberg- and convoy electrons, we can hardly see how the question of convoy electron production could be solved without tackling the formation of Rydberg states in the very same collision systems.

Formation of Rydberg-Ions behind Foil Targets

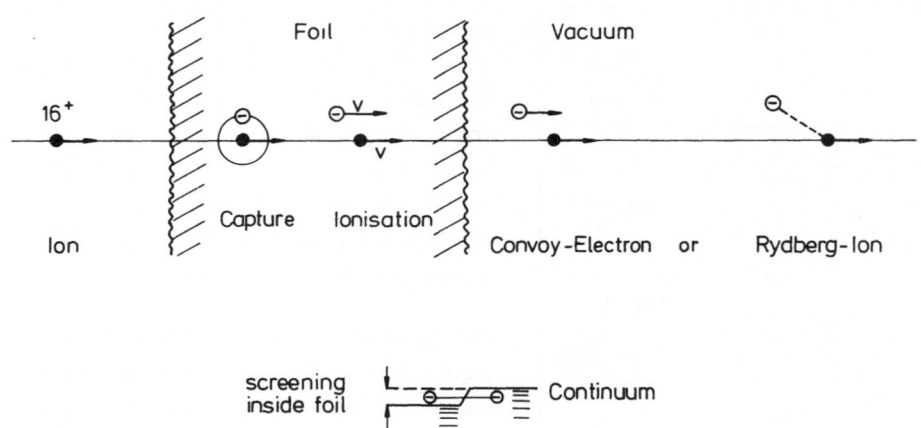

Fig.8: Hypothetical multi-step model for the formation of Rydberg states of fast ions emerging from solid targets.

V. Quenching of Rydberg States

Our conclusions (i) and (ii) in Section III can be put to a further test. We applied a transverse electrical field across the beam in the viewing region of the x-ray detector and searched for changes in the x-ray spectra and in the decay curves. Again, gas- and foil targets entailed completely different effects. Gas-excited ions caused no detectable difference at all, but foil-excited beams showed a dramatic rise of some x-ray lines (Fig.10 and 11). While the Ly-α line exhibits only a modest increase with increasing strength of the quenching field, the $np \to 1s$ transitions increase dramatically in intensity especially for large n. In fact, the two strong lines near 3 keV in Fig.10c are centered around the series limit of hydrogen- and helium-like sulfur ions, lines which are practically not detectable without application of an external field.

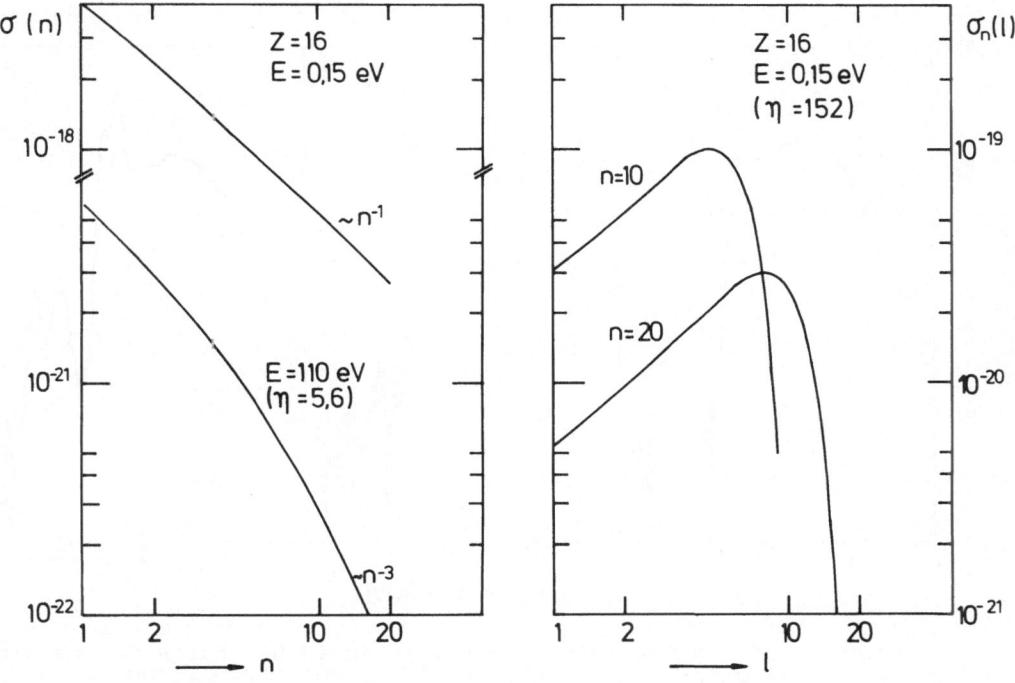

Fig.9: Cross section in cm² for radiative electron capture (REC) of free electrons by bare sulfur ions. E denotes the kinetic energy $mv^2/2$ of the electron relative to the ion, η stands for the Sommerfeld parameter $v_0 Z/v$ ($v_0 = e^2/\hbar$). The left part displays the dependence on the final principal quantum number, whereas the right part illustrates the resulting ℓ-distribution.

These results must be regarded as qualitative confirmation of our previous conclusions about ℓ-distributions. Let us first consider Rydberg states formed by electron capture in ion-atom encounters: as we have illustrated before, low-ℓ states are predominantly populated, i.e. states which have the shortest lifetimes within a principal shell. Therefore, admixture of high-ℓ states offers no chance for increased decay probabilities and, thus, we detect no enhanced x-ray intensities. In contrast, foil-excited Rydberg states are assumed to have high angular momentum, i.e. lifetimes are much longer and Stark mixing offers the chance for much faster decays via, say, the corresponding np-level. As a consequence, we detect np → 1s transitions for very high n-values. In a rough analytical approximation of this quenching model we derive $I(E) \propto E^{5/4}$ which is not too far off from the actually observed dependence I ∝ E for these transitions (E stands for the field strength).

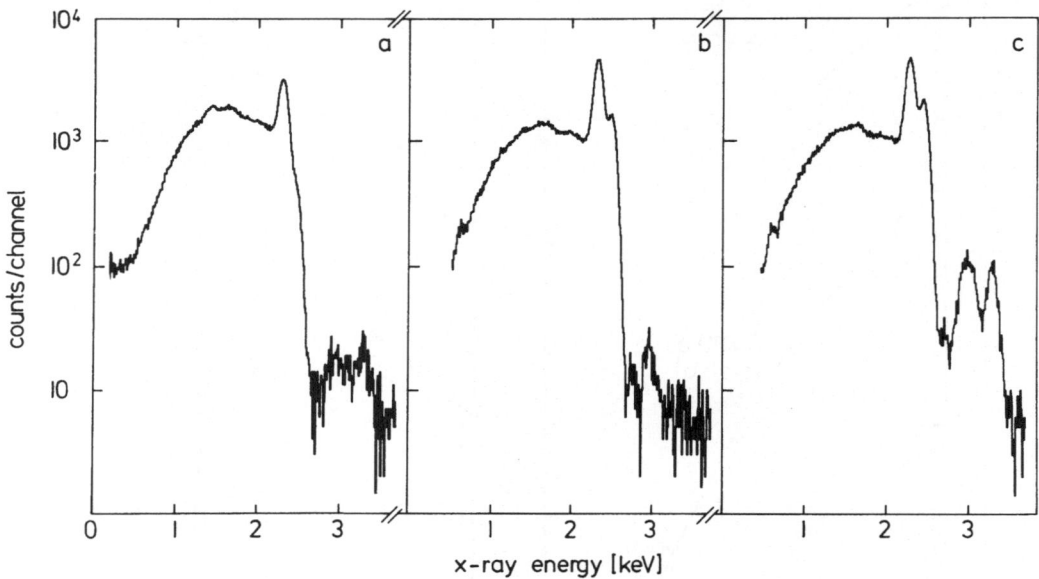

Fig.10: K x-rays of 125-MeV sulfur ions with incident charge 16⁺ passing through (a) 3×10^{16} molecules N_2/cm^2, (b) 10 µg/cm² carbon, and (c) 10 µg/cm² carbon with an external electrical field of 3 KV/cm across the observation region. In all cases, the distance between target and detector amounted to 7.5 cm. For line identification see text and previous figures. Note the spectral differences in the Kβ region near 3 keV.

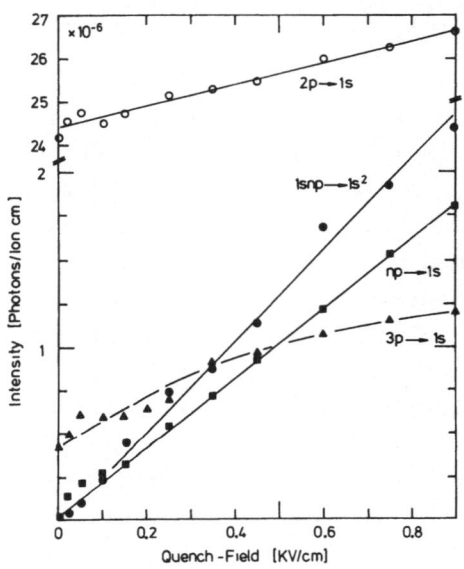

Fig.11 Intensity of K x-ray transitions shown in Fig.10c, as a function of quench-field intensity. The lines 1snp → 1s² and np → 1s occur at the series limit of helium- and hydrogen-like K-shell transitions and, thus, indicate decays with large values of n.

Again, a more quantitative reproduction of this experimental result is
hard to achieve. Stark mixing in Rydberg states with the inclusion of
all possible ℓ-m substates and a variety of n-states with binding energies both small, comparable and large compared with Stark shifts represents a problem of computationally formidable complexity and has not
yet been tackled successfully even for 1-electron ions.

VI. Conclusion

In recent years extensive studies of convoy electrons have been presented which revealed a number of interesting features. It has beccme clear,
for example, that capture to the continuum (ECC) can be singled out experimentally and can be understood theoretically to an impressive extent. In ion-foil interaction, however, ECC is not expected to represent
a dominant mode because the required large momentum transfer to the captured electron renders the cross section relatively small. As soon as
projectile electrons are available, especially when they reside in states with low binding energy, loss to the continuum (ELC) becomes a very
likely process. Nevertheless, basic features of ELC remained unclear.

We have studied formation of Rydberg states and found a number of surprising effects; the actual mode of Rydberg-state formation remained
unexplained, but we present good reasons to view it as a multi-step process closely linked to ELC. Among other possible tests of our conjec-
tures it would certainly be interesting to see whether convoy yields
exhibit the same target-thickness dependence as was found for Rydberg
states, provided the required conditions can be met such as pre-equi-
librium with respect to charge exchange inside the solid, well defined
incident charges states, and correlation of electrons with particular
ionic charge states.

We expect that measurements of convoy electrons and observation of Rydberg states by either x-ray techniques or field ionization methods produce converging results and contribute to a satisfactory explanation
of the basic collision phenomena ocurring in ion-foil interaction.

This work was partially supported by the Bundesministerium für Forschung
und Technologie.

References

1. P.Richard, Phys.Lett. 45A, 13 (1973).

2. W.J.Braithwaite, D.L.Matthews and C.F.Moore, Phys.Rev. A11, 465 (1975).

3. R.M.Schectman, Phys.Rev. A12, 1717 (1975).

4. L.J.Curtis, R.M.Schectman, J.L.Kohl, D.A. Chojnacki and D.R.Shoffstall, Nucl.Instr. and Meth. 90, 207 (1970).

5. L.J.Curtis, Am.J.Phys. 36, 1123 (1968).

6. H.-D.Betz, J.Rothermel, D.Röschenthaler, F.Bell, R.Schuch and G.Nolte, Phys.Lett. 91A, 12 (1982).

7. J.Rothermel, H.-D.Betz, F.Bell and V.Zacek, Nucl.Instr. and Meth. 194, 341 (1982).

8. J.E.Bayfield, G.A.Khayrallah and P.M.Koch, Phys.Rev. A9, 209 (1974).

9. H.-D.Betz, D.Röschenthaler and J.Rothermel, Phys.Rev.Letters 50, 34 (1983).

10. M.Breinig, S.B.Elston, S.Huldt, L.Liljeby, C.R.Vane, S.D.Berry, G.A.Glass, M.Schauer, I.A.Sellin, G.D.Alton, S.Datz, S.Overbury, R.Laubert and M.Suter, Phys. Rev. A25, 3015 (1982).

11. R.W.Hasse, H.-D.Betz and F.Bell, J.Phys. B12, L711 (1979).

12. H.-D.Betz, J.Rothermel and F.Bell, Nucl.Instr. and Meth. 170, 243 (1980).

13. L.Dubé, J.Phys. B17, 641 (1984).

14. D.Belkić, R.Gayet and A.Salin, Comp.Phys.Comm. 30, 193 (1983).

CONVOY ELECTRONS FROM ATOMIC AND MOLECULAR HEAVY

ION COLLISIONS WITH SOLIDS[§]

P.Koschar, R.Latz, J.Kemmler, M.Burkhard, H.J.Frischkorn, D.Hofmann,
J.Schader, R.Schramm, K.O.Groeneveld

Institut für Kernphysik, J.W.Goethe Universität Frankfurt/Main,
August Euler Straße 6, 6000 Frankfurt/Main, Germany

M.Breinig, S.Elston, I.A.Sellin

University of Tennessee, Knoxville and
Oak Ridge National Laboratory, Oak Ridge, Tennessee, USA

and

W.Meckbach

Centro Atómico, Bariloche, Argentina

Abstract

Energy distributions of convoy electrons have been measured for different projectiles (0.03 MeV/u < E_p/M_{p_2} < 2.7 MeV/u) traversing thin carbon foils (2 $\mu g/cm^2$ < ρx < 20 $\mu g/cm^2$). The motivation of our study
is to elucidate the mechanisms of convoy electron production in ion-
solid collisions. The analysis of the convoy peak's shape and yield
dependence on dwell time t_d, velocity v_p, incident charge state q_i and
atomic number z_p of the projectile shows significant deviations from
theories of ECC and ELC in single collision systems. In non-equilibrium
of the projectile charge (t_d < 2 fs) the convoy electron yield Y_c of
atomic heavy ions depends strongly on q_i and exhibits an exponential
dependence on the dwell time. Here the ELC process prevails. In the
higher dwell time regime (t_d > 2 fs) the convoy electron yield Y_c is
described by $Y_c \sim z_p^{3.2} E_p^{-1.3}$. In comparison to free isotachic electrons the convoy electrons have an effective mean free path $<\lambda_c>$ about
two up to ten times higher. The value $<\lambda_c>$ depends on the velocity v_p
and the nuclear charge z_p and suggests a strong correlation between the
convoy electron and the traversing ion. We discuss the observed convoy
production and their strong dwell time dependence (t_d < 30 fs).

1. Introduction

In the course of the last decades ion-solid interaction mechanisms were
thoroughly investigated, and, as an extremely successful technique, applied in beam-foil spectroscopy. On the one hand the study of ion-solid
interaction sheds light on the excitation and ionisation processes and
the decay of electronic states of projectile and target atoms, on the
other hand it reveals aspects of collective phenomena in condensed matter first discussed in the macroscopic quantities of energy loss and
angular straggling of projectiles in solids [1]. The discovery of convoy electrons by K. G. Harrison and M. W. Lucas [2] was the beginning of
a new approach to this area of research.
Convoy electrons are particularly appropriate to expose the history of
the projectile ion and its environment inside the solid because the velocity v_e of these electrons matches the velocity v_p of the emergent
ion both in speed and in direction [3]. From measurements with gaseous
targets we know that these electrons originate from capture to the continuum (ECC) for bare or nearly bare projectiles [4], and from loss
to the continuum (ELC) when loosely bound projectile electrons are
available [5]. In the case of solids further processes are discussed
which take into consideration the properties of condensed matter:

(1) The indirect electron loss to the continuum (IELC) [6]:
During the passage the ion possibly captures one electron into
a bound state and in a second step looses the electron by the
ELC process.

(2) Wake riding electrons (WRE) originate from bound states centered in the electron density minima of the wake potential induced by the penetrating ion charge through the free electron
plasma of the solid [7].

(3) Field ionisation of projectile Rydberg states during the passage
through the target surface [8], [9].

Extremely useful objects in beam foil spectroscopy are diatomic, molecular projectile ions. After the molecular ion break-up in the entrance
layer of the solid, the molecular fragments disintegrate via Coulomb
explosion [10]. The motion is not one of independent fragments but of
particles correlated in space and time. Each fragment is trailed by an
oscillating, damped electron density fluctuation which can be related
to its potential Φ, the "wake potential". Their wake potentials are
superimposed and can modify dramatically many observable quantities
such as the transmission and reconstitution probabilities of molecular

ions along their trajectories through solids [11], the electron loss or capture to continuum states [6], [12], [13], the stopping power [14], the joint centre of mass, angular and energy distribution of the fragments [15] and the total electron yield [16].

It is the aim of this paper to summarize recent experimental results of convoy electron production with atomic and molecular ions.

2. Experiment and data reduction

The experimental arrangement is schematically shown in fig. 1. We used atomic and molecular projectile ions H^+, He^+, He^{++}, C^{4+}, C^{6+}, H_2^+, H_3^+ (specific energy 0.8 MeV/u < E_p/M_p < 2.7 MeV/u) and C^+, N^+, O^+, O^{++}, CO^+, N_2^+ (0.03 MeV/u < E_p/M_p < 0.07 MeV/u) from the Frankfurt University 7.5 MV and 2.5 MV Van de Graaff accelerators and the atomic heavy ions Ni^{+24}, Ni^{+26}, Ti^{+14} and U^{+44} with 1.4 MeV/u from the UNILAC of the GSI Darmstadt. The beam was collimated to 0.6 mm diameter and 0.3^0 angular spread and penetrated either poly-crystalline carbon foils (2 $\mu g/cm^2$ < ρx < 25 $\mu g/cm^2$) or a dynamic CH_4 - gas target used for measurements under single collision conditions. The foil or gas target was easily exchangeable without altering other experimental conditions. This permitted a direct comparison of yields and peak shapes produced in ion-atom and ion-solid interaction.

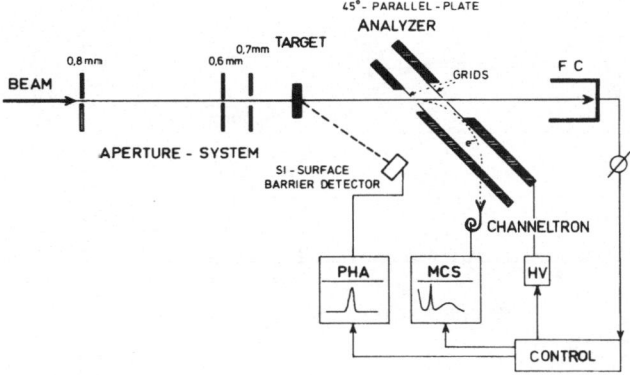

Fig. 1. Schematic diagram of the experimental apparatus with the 45^0 - parallel plate electron analyzer.

The thicknesses of the targets were monitored during the experiment by elastically scattered particles with a silicon surface barrier detector positioned at 35° to the beam direction. The elastic scattering cross sections were assumed to be the same for atomic projectiles and molecular fragments of the same species [17]. The actual foil thickness ρx was calculated from the elastic scattering yield normalized to the one of carbon foils of known thickness (here: $\rho x = (22 \pm 1)\mu g/cm^2$, measured by transmission of infrared radiation) [18]. The ρx-values were transformed to the dwell time t_d of the projectile inside the solid by $t_d = x/v_p$, assuming a mass density of carbon of $\rho = 1.9 \ g/cm^3$ [19].

The particle beam passed with almost no parasitic scattering through the electron analyzer system and was stopped in a Faraday cup. The particle and electron spectra were normalized to the charge accumulated in the Faraday cup. Zero degree electron spectra were measured either at electron energies $E_e > 100$ eV with a 90° magnet analyzer [20] or at $E_e > 5$ eV with a modified 45° parallel plate electrostatic analyzer. Both analyzing systems provided a good energy resolution of $\Delta E_e/E_e = 1.3 \pm 0.1\%$ with an acceptance angle of $\Delta\Theta = 1.1°$, set by the beam diameter at the position of the target and the analyzer exit aperture.

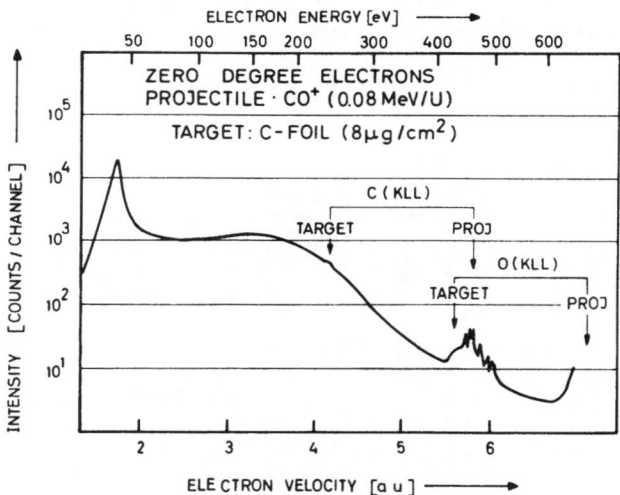

Fig. 2. Zero degree electron spectrum (transmission uncorrected) from CO^+ projectile (0.08 MeV/u) traversing a thin carbon foil ($\rho x = 8.0 \ \mu g/cm^2$). Auger electron peaks from target and projectile are marked. Note the strong (log. scale!) cusp-peak at an electron energy of 44 eV.

The measurements in the low energy range E_e < 50 eV required a sufficient shielding of the magnetic stray fields down to $5 \cdot 10^{-6}$T. The described experimental arrangement also provides the data of free monoenergetic electrons transmitted through carbon foils. Here, the projectile beam is replaced by an electron beam produced in an electron gun. This permits, without changing the experimental set-up, the direct comparison between convoy electron spectra induced by projectile ions and the energy distribution of free electrons after traversing the solid.

A typical electron velocity spectrum from molecular CO^+ - projectiles is shown in fig. 2. Beside the binary encounter (BE) electrons and target and projectile Auger peaks (marked in the figure) the dominant structure is the cusp shaped convoy electron peak at $v_e = v_p$.
Fig. 3 gives another example of a zero degree electron velocity spectrum from heavy projectile $^{238}U^{+44}$ (1.4 MeV/u) traversing a thin carbon foil.
For the purpose of yield determination the spectra are corrected with the analyzer transmission measured with a monoenergetic, intensity stabilized electron gun.

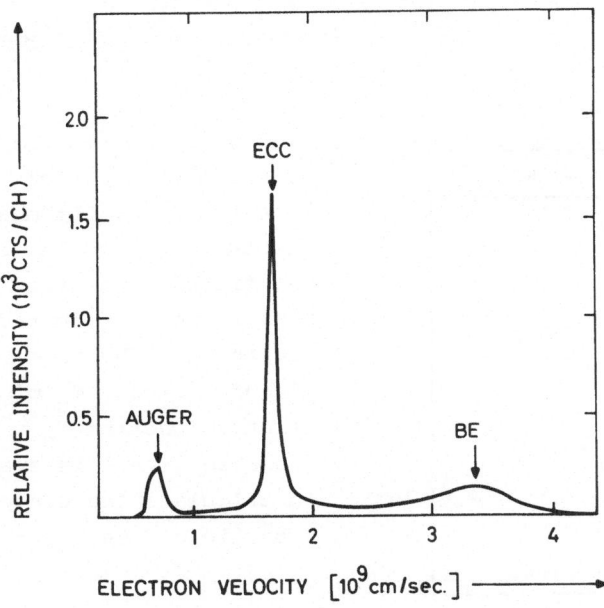

Fig. 3.   Zero degree electron spectrum from a carbon foil ($\rho x = 20$ $\mu g/cm^2$) bombarded by $^{238}U^{44+}$ ions (1.4 MeV/u).

The background in the energy range of the convoy electron peak is mainly composed of projectile ionisation electrons and secondary electrons emitted from the solid.

In single collision conditions we can not distinguish between ionisation electrons ejected under zero degree and ECC-electrons. The quantum mechanical description shows that the cross sections for direct ionisation and the one for electron capture to the continuum cannot be calculated independently. Ionisation electrons ejected from solids predominantly originate from angles unequal to zero degree and are detected under zero degree because additionally they are scattered into the spectrometer acceptance angle. Here both the contribution of ionisation electrons and secondary electrons can be subtracted from the convoy peak [21].

Fig. 4. shows a measured spectrum I_T (solid line) and the fitted background I_u (dotted line) evaluated between broad limits using the function $I = A \cdot v_e^B$:

- for light atomic projectile ions, $v_p - 2v_B < v_e < v_p + 2v_B$
- for light molecular projectile ions, $v_p - 6v_B < v_e < v_p + 4v_B$

    both with high specific energies 0.8 MeV/u $< E_p/M_p <$ 2.7 MeV/u and

    in the case of heavy ions, $0.7\, v_p < v_e < 1.3\, v_p$

$$(0.03 \text{ MeV/u} < E_p/M_p < 0.07 \text{ MeV/u}).$$

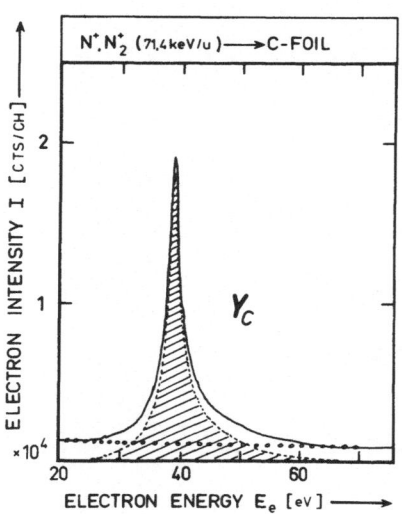

Fig. 4. Reduction of the convoy electron yield Y_c from measured data: The measured spectrum I_T (solid line) and the fitted ionisation electron background I_u (dotted line) are shown. The total area (shaded area) N_c of the difference spectrum $N_c = I_T - I_u$ (dashed line) is evaluated to calculate the convoy electron yield (see text).

The limits are determined by comparison with PWBA - calculations and measurements with gaseous targets. Thus we can determine the integrated number of counts N_C (shaded area) of the difference spectrum $I_C = I_T - I_u$ (dashed line). The yield Y_C of convoy electrons per projectile and steradian is obtained from N_C using the relation:

$$Y_C = \frac{N_C}{N_p \cdot \varepsilon \cdot \Delta\Omega}$$

(N_p: number of projectiles, ε: spectrometer efficiency, $\Delta\Omega$: solid angle). Instead of measuring the quantities ε and $\Delta\Omega$ independently, their product can be derived from the yield of carbon KLL - Auger electrons from a CH_4 - gas target bombarded by H^+ (1.7 MeV). The data are normalized to tabulated cross sections from ref. [22]. The uncertainties in the absolute values of Y_C are estimated to be ±30% mainly caused by the uncertainty of the absolute cross sections from ref. [22].

The number of projectiles N_p is calculated from the measured Faraday cup charge Q divided by the mean charge $\bar{q}$ of the particle after penetrating the target. The mean charge $\bar{q}$ for specific energies $E_p/M_p >$ 0.8 MeV/u are tabulated in ref. [23]. In the case of $E_p/M_p <$ 0.1 MeV/u we measured the mean charge $\bar{q}$ for the projectiles N^+, C^+, O^+, N_2^+, CO^+ under charge equilibrium conditions after traversing the target [24].

3. Results and discussion

To distinguish different mechanisms of convoy electron production we use either bare projectile ions (e.g. H^+) or ions with initially bound electrons (e.g. H_2^+). Under single collision conditions the ELC cross section usually is much larger than the ECC cross section; therefore, a similar cross section ratio for these two processes in ion - solid collision can be expected. In fig. 5 the convoy electron yields $Y_C(H^+)$ and $Y_C(H_2^+)$ are plotted as a function of the specific target thickness ρx at different specific projectile energies E_p/M_p.

The yields Y_C for bare ions approaches a saturation value Y_{CS} with increasing target thickness ρx. The energy dependent data can be fitted by a power function

$$Y_{CS} = a \cdot E_p^{-m} \qquad \text{with} \quad m = 1.3 \pm 0.1$$

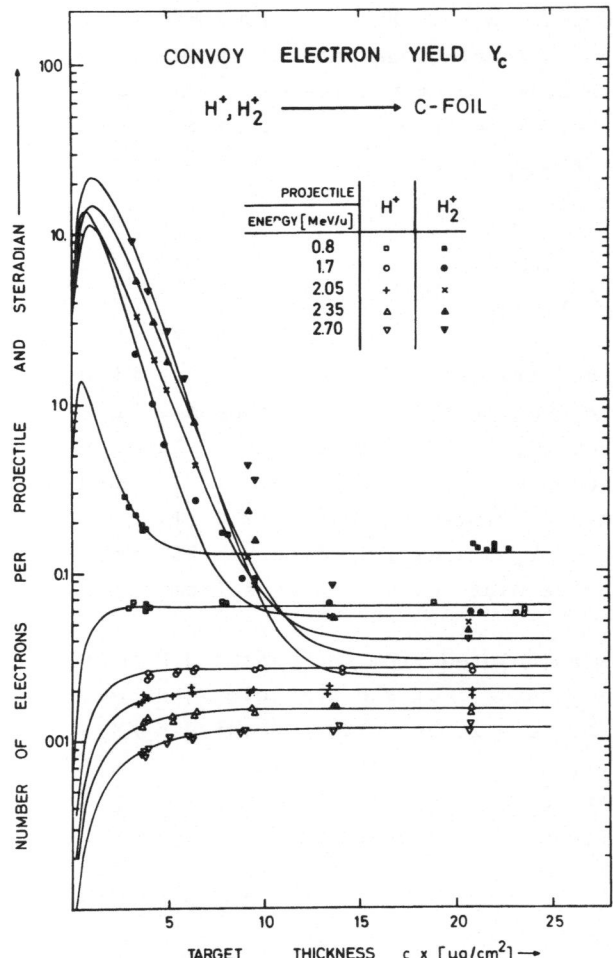

Fig. 5

Convoy electron yields Y_C (see fig. 3) from H^+ and H_2^+ projectiles as a function of the specific target thickness ρx at different specific projectile energies E_p/M_p. The curves represent fits of the convoy electron yield (see text).

for H^+ and He^{++} (0.8 MeV/u $< E_p/M_p <$ 2.7 MeV/u).

Under the assumption that only the ECC process and the electron scattering in solids contribute to the convoy electron yield of bare nuclei, the thickness dependence of Y_C is given by

$$Y_C(E_e, x) = A \times (1 - \exp (- \mu_e \times x)).$$

Here, the ratio $A = [d \sigma_+(E_e)/d\Omega]/ \sigma_-$ is given by the single differential cross section for electron capture $d\sigma_+(E_e)/d\Omega$ and the total electron scattering cross section σ_- ; the electron scattering coefficient μ_e, the density of target nuclei n_T and σ_- are connected by $\mu_e = n_T \times \sigma_-$ (compare e.g. ref. [12]). The value μ_e relates to the mean free path $<\lambda_C>$ of a convoy electron inside the solid by $<\lambda_C> = \mu_e^{-1}$.

Table 1. Single differential cross sections $d\sigma_+/d\Omega$ for electron capture, total electron loss cross sections σ_- and electron mean free paths $\langle\lambda_c\rangle$ and $\langle\lambda_a\rangle$ measured with projectile ions and free electrons traversing carbon foils.

projectile	E_e [eV]	E_p/M_p [MeV/u]	$d\sigma_+/d\Omega$ [10^{-20}cm^2/sr]	σ_- [10^{-18}cm^2]	$\langle\lambda_c\rangle$ [Å]	$\langle\lambda_a\rangle$ [Å]
H^+	436	0.8	170 ± 15	26.7 ± 2.5	43 ± 4	
	932	1.7	51.2 ± 5	18.7 ± 1.8	62 ± 6	
	1116	2.05	22.5 ± 2	11.3 ± 1.0	100 ± 9	
	1279	2.35	13.6 ± 0.9	8.8 ± 0.9	129 ± 13	
	1470	2.7	8.56 ± 0.7	7.1 ± 0.7	160 ± 15	
D^+	932	1.7	49.1 ± 4.5	18.2 ± 1.8	62 ± 6	
$^3He^{++}$	932	1.7	391 ± 35	14.7 ± 1.5	77 ± 8	
Ni^{+24}, Ni^{+26}	8330	15.2	–	–	2000	
free e^-	272	0.5		93.0 ± 6.6		7.0 ± 0.5
	932	1.7		26.7 ± 0.5		24.5 ± 0.5
	1580	2.9		12.0 ± 0.3		53.3 ± 1.5
	2100	3.86		9.5 ± 0.2		68.4 ± 1.0
H^+ (CH_4-gas target)	932	1.7	40 ± 2			

The solid lines for H^+ in fig. 5 represent fits with the procedure described above. Tab. 1 gives deduced values of $d\sigma_+/d\Omega$, σ_- and $\langle\lambda_c\rangle$ of the present experiment. Also, scattering cross sections derived from measurements with monoenergetic electrons penetrating carbon foils are included.

The single differential cross section for electron capture depends on the nuclear charge z_p and the energy E_p of the projectile as

$$d\sigma_+ / d\Omega \sim z_p^n \times E_p^{-m} .$$

Breinig et.al. [6] report on experimental values of $n = 2.75 \pm 0.2$ and $m = 2.25 \pm 0.1$ at high z_p and at high projectile velocities (1 MeV/u < E_p/M_p < 4 MeV/u) and from few H^+ - data lower m - values. From the

Table 2. $z_p^n \times E_p^{-m}$ - dependence of $d\sigma_+/d\Omega$

projectile nuclear charge z_p	target nuclear charge z_t	$v_p^2 \pm \Delta v_p^2$ [MeV/u]	n	m	Ref.
		- solid target -			
1,8,14,28	6,13,47,79	2.5 ± 1.5	2.75 ± 0.2	2.25 ± 0.1	[6]
1,2	6	1.7 ± 1.0	2.9 ± 0.2	1.6 ± 0.2	this exp.
1	6	0.35 ± 0.15	-	3.1	[25]
1	6	1.5 <	3.2 ± 0.2	-	[26]
		- gas target -			
1,6,8,14	18	2.5	2.3 ± 0.3	-	[6]
6,8,14	2	2.85 ± 1.0	-	4.2 ± 0.3	[6]
6,8,14	18	2.85 ± 1.0	-	2.15 ± 0.15	[6]
1	10	0.75 ± 0.45	-	4.8	[27]
1	H_2	0.75 ± 0.75	-	4.75	[28]

data of this experiment which includes the projectiles H^+, D^+, $^3He^{++}$, $^4He^{++}$ we find n= 2.9 ± 0.2 and m= 1.6 ± 0.2 . Measurements of Meckbach et.al. [25] for H^+ in the energy range of 0.2 - 0.5 MeV give a $E_p^{-3.1}$-dependence. Tab. 2 summarizes n- and m- values for gaseous and solid targets. Only for single collisions with targets of low z_t - values the energy dependence agrees with Dettmann's m = 5 prediction. Here, the orbital velocity of target electrons is much smaller than the projectile velocity v_p and mainly s - electrons can contribute to ECC as assumed in the Born approximation. The strong deviation between calculated and experimental energy dependencies of Y_C indicates other mechanisms in the ion-solid interaction.

One explanation could be the dynamic screening of the nuclear charge in the solid by the electron plasma byond the screening length $a_s = v_p/\omega_{pl}$ (ω_{pl} plasma frequency). This leads to a weaker decrease of $d\sigma_+/d\Omega$ with increasing E_p.

In addition to the ECC process, the radiative electron capture into continuum states (RECC) in the higher projectile energy regime ($v_p \gg v_B$) is of interest [29],[30]. Loosly bound or free electrons, available mainly in the electron plasma of the solid, can be captured into projectile continuum states only while emitting photons. The analysis gives symmetric cusp peaks and predicts a weak $E_p^{-2.5}$ - dependence. Compared to the ECC cross section the RECC process dominates only for $v_p > 23\, v_B$ ($z_t = 6$). Under our experimental conditions the contribution of RECC is in the order of 1% .

In the velocity regime of this experiment the charge state distribution of projectile ions reaches an equilibrium in the first layers of the solid. A fraction of projectile ions captures electrons and is then able to contribute to the convoy electron production by the ELC - process. This two step mechanism, refered to as " indirect electron loss to continuum states (IELC) " , depends on the charge state fractions of projectile ions with captured electrons F ($q \neq z_p$) and the ELC cross sections $d\sigma(q)/d\Omega \lceil^{ELC}$ as function of the charge state q. In the case of H^+ (1.7 MeV/u) we only have to consider the fraction F (0) of H^O. The fraction F (0) depends on the ratio

$$\sigma_C(H^+) \; / \; [\sigma_C(H^+) + \sigma_L(H^O)]$$

of the total electron capture cross section $\sigma_C(H^+)$ and the sum of $\sigma_C(H^+)$ and the total electron loss cross section $\sigma_L(H^O)$. Here, the IELC cross section σ_{IELC} is given by σ_{IELC} = F (0) × $d\sigma(H^O)/d\Omega\lceil^{ELC}$ and can be estimated using values for σ_L and σ_C of 3.1 × 10^{-17}cm^2 and 2.1 × 10^{-21}cm^2, respectively; they are deduced from transmission measurements of H^+ and H^O through carbon foils [31] and from transmission measurements of H_2^+ - projectiles (instead of H^O) traversing CH_4 - gas targets ($d\sigma(H_2^+)/d\Omega\lceil^{ELC}$ = (126 × 10) × 10^{-17}cm^2/sr). The comparison of the IELC - cross section σ_{IELC} = 8.5 × 10^{-20}cm^2/sr, thus calculated, with the ECC - cross section $d\sigma_+(H^+)/d\Omega$ = (40 × 2) × 10^{-20}cm^2/sr (see tab. 1), measured under single collision condition, shows that the IELC mechanism possibly enhances the convoy electron yield to 21% and may account for the difference between the $d\sigma_+/d\Omega$ values in solid and gas targets.

The ELC - cross section, calculated in the frame of the PWBA [32], depends only weakly on the projectile energy ($\sim E_p^{-m}$, m < 1) and approaches a constant value for $v_p \gg v_B$. This can explain the strong deviations between the energy dependence of Y_C under single collision conditions and the energy dependence of Y_C in collision with solids.

The measured dependence of Y_{CS} on the nuclear charge z_p is in good agreement with the predicted z_p^3 - scaling law [33]. For heavy ions the value n = 2.75 ± 0.2 has been shown to be independent on E_p [6]. From the present data we find a slight increase of n = 3.2 ± 0.2 (E_p/M_p = 1.7 MeV/u) with increasing projectile energy E_p.

The projectile nuclear charge z_p not only influences the production of

convoy electrons but possibly also changes the scattering of these
electrons inside the solid. The attractive Coulomb interaction between
the projectile and the correlated electron is proportional to the effective charge of the ion and may course a refocussing of the electron
back to the projectile. It may reduce the multiple scattering [34].
This so called " Coulomb focussing " is discussed in more detail by
Sellin et.al. in this volume [35].

The yield Y_C and, especially, the deduced mean free path $<\lambda_C>$ may re-
flect this phenomenon too [34]. The interpretation of this quantity
$<\lambda_C>$ is questionable, because within the frame of this model Y_{CS} is
only sensitive to the ratio of [$d\sigma_+/d\Omega$]/ σ_- . But, for small target
thicknesses the ρx - dependence of Y_C will be described only with the
electron scattering cross section σ_- or $<\lambda_C>$, respectively (see tab.1).
Fig. 5 indicates an enhancement of Y_C (H^+) of 50% with increasing
target thickness (note the logarithmic scale).

Actually, $<\lambda_C>$ - values of convoy electrons produced by light ions are
up to three times higher than those $<\lambda_a>$ measured for free isotachic
electrons. From experiments with swift heavy ions, e.g. Ni^{+24}, Ni^{+26}
(15 MeV/u), we found an even higher value of $<\lambda_C>/<\lambda_a>$ = 20 [35].
In addition, the comparison of the ratio $<\lambda_C(He^{++})>/<\lambda_C(H^+)>$ of isotachic He^{++} and H^+ shows a weak increase with increasing z_p. This also
indicates a focussing mechanism.

In the present model further processes in convoy electron production,
like IELC, are neglected. This has to be considered in determining
$<\lambda_C>$ values. Here, measurements with very thin targets ($\rho x < 2$ $\mu g/cm^2$)
or higher velocities may give more detailed information.
The yield curves of incoming projectiles with bound electrons show a
very different dependence on the target thickness ρx (see fig.5).
For small ρx - values ($\rho x < 10$ $\mu g/cm$) called the " nonequilibrium
regime " or " red regime " [36], the yield curves decrease exponentially,
again approaching a constant value Y_{CS} at larger target thicknesses,
called the "equilibrium regime" or " blue regime " [36]. The exponential
decrease indicates a large fraction of electrons, lost from initially
bound states of the projectile into the continuum, which contribute to
the convoy electron yield Y_C. This explanation is supported by $Y_C(\rho x)$ -
measurements with isotachic O^+ - and O^{++} - projectiles (0.04 MeV/u <
$E_p/M_p < 0.07$ MeV/u); they result in a very weak dependence on the
initially charge states q_i only in the nonequilibrium regime
($R = Y_C(O^+) / Y_C(O^{++})$ = 1.1 ± 0.05).

We may decribe the ELC - contribution Y_L to the total yield Y_C in the

nonequilibrium regime in a similar way as outlined for bare projectiles:
1. instead of capturing an electron the projectile keeps it's electron
with the probability $\exp(-n_T \times \sigma_L \times l)$, where σ_L is the total cross
section for electron loss in solids until it reaches the depth l ;
2. the electron is emitted in the depth l into the forward direction
given by the single differential cross section $d\sigma_L(\theta=0)/d\Omega$:
3. the electron is scattered in the remaining layers $(x - l)$ inside
the solid described by μ_e. Thus the yield Y_L is expressed by

$$ Y_L = \frac{d\sigma_L(\theta=0)}{d\Omega} \times \frac{n_T}{(\mu_e - n_T \times \sigma_L)} \times [\exp(-n_T \times \sigma_L \times x) - \exp(-\mu_e \times x)] $$

[12].

Because in the equilibrium regime the fraction of Y_L vanishes it can
not contribute to the saturation yield Y_{CS}. Here, the initial charge
state q_i of the incoming ion has approached the dynamic equilibrium
charge state distribution independent on q_i. We assume that all
processes ECC, ELC and IELC contribute to Y_{CS} in dependence on the
equilibrium charge state fractions F $(q=z_p)$ and F $(q \neq z_p)$, respectively.
Thus the convoy electron yield Y_{CS} is independent on the initial charge
state q_i. The value Y_{CS} will be described in terms of an extended model
including ECC, ELC and IELC mechanisms.

Studies with swift He^{++}, He^+ (1.7 MeV/u) and C^{+2}, C^{+6} (1 MeV/u),
and slow O^{++}, O^+ (0.04 MeV/u $< E_p/M_p <$ 0.08 MeV/u) projectiles indicate no q_i - dependence of the total yield Y_{CS}. Also, Y_C of swift heavy
projectile ions with initial charge states q_i nearly equal to the mean
charge $\bar{q}$ in solids, e.g. Ti^{+14} and $^{238}U^{+44}$ (1.4 MeV/u), are independent on ρx indicating the relation to the charge state equilibrium.

Further evidence of initial charge q_i - independence gives the angular-
and energy- distribution of Y_C. Fig. 6 shows an example of the angular
distribution of Y_C measured with C^{+2} and C^{+6} projectiles penetrating
a thin carbon foil ($\rho x = 3 \mu g/cm^2$). We conclude: Within the frame
of these experimental conditions, the production of convoy electrons
is independent on q_i.

Considering these experimental results we can describe the $Y_C(\rho x)$
yield curves of projectile ions with initially bound electrons, e.g.
H_2^+, with the sum of Y_L and Y_{CS}, respectively (see fig. 5). Ref. [12]
summarizes the deduced total loss cross sections σ_L and the single
differential cross sections $d\sigma_L(\theta=0)/d\Omega$ for different light atomic
and molecular projectile ions. Good agreement is found e.g. for He^+

between our experimental cross section

$$\sigma_{L,exp} = (1.5 \pm 0.4) \times 10^{-17} cm^2$$

and the calculated cross section

$$\sigma_{L,calc} = 1.7 \times 10^{-17} cm^2$$

using Bohr's relationship [1]. This good agreement contrasts to the experimental cross section

$$\sigma_{L,trans} = (3.96 \pm 0.3) \times 10^{-17} cm^2$$

deduced from charge state q=1 transmission measurements of incident He$^+$ [37].

The discussion of the $Y_C(\rho x)$ yield curves reveals contributions of different convoy electron production mechanisms. Further insight in these mechanisms promisses the analysis of the convoy electron peak's shape i.e. it's longitudinal and transversal velocity distribution ($v_{e\parallel}$ and $v_{e\perp}$) [3]. Let us pay attention to the analysis of the energy distribution of ELC - electrons ($Y(v_\parallel)$ at $\theta = 0^o$).

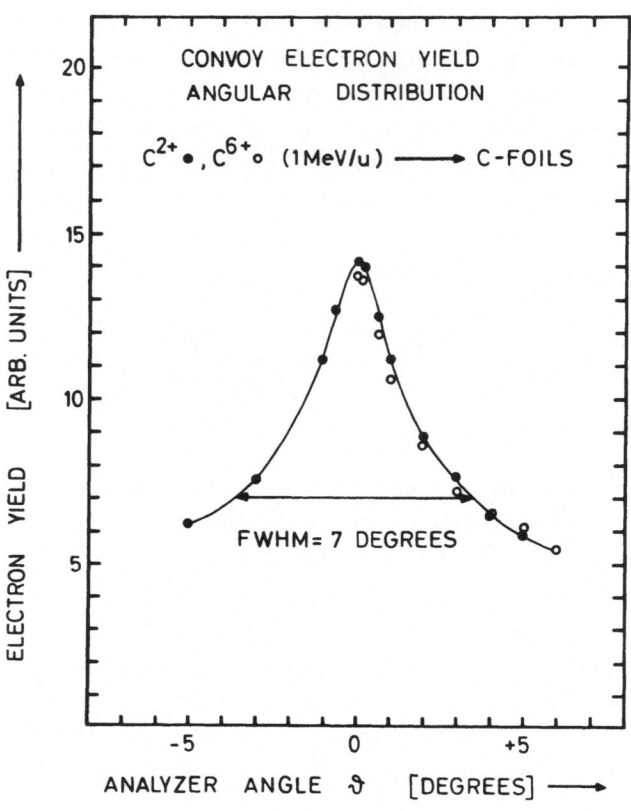

Fig.6. Angular distribution of the convoy electron yield Y_C (see fig.3) measured with isotachic C^{+2} - and C^{+6} - projectiles (1.0 MeV/u) traversing a thin carbon foil ($\rho x = 3$ $\mu g/cm^2$).

In the nonequilibrium regime the ρx - dependence of the yield Y_C and the energy distribution of convoy electrons are determined by ELC electrons (typical examples are published in fig.2 of ref. [20] or in fig.1 of ref. [38]). The energy distribution can be calculated in the three steps which had been used to represent the $Y_C(\rho x)$ curves (see above): 1. The projectile ion with an initially bound electron keeps it's electron in the depth x' according to the probability exp ($-\sigma_- \times n_T \times x'$); 2. the electron emerges with a typical cusp shaped energy and angular distribution $d^2\sigma$ ($\theta = 0$)$/d\Omega dE_e$;3. this electron is scattered according to σ_- and suffers energy loss, energy- and angular-straggling in the remaining layers (x-x') of the solid target. The distribution at the end of the foil can be evaluated by numerical integration of the contribution of each infinitesimal target layer x' between x' and (x' + dx). Two preceding experiments are nessesary to deduce experimental values for our semiempirical model.

1. Transmission measurements of free monoenergetic electrons passing through carbon foils provided numerical, experimental data of both the scattering coefficient μ_e and the electron energy distribution as a function of ρx and initial

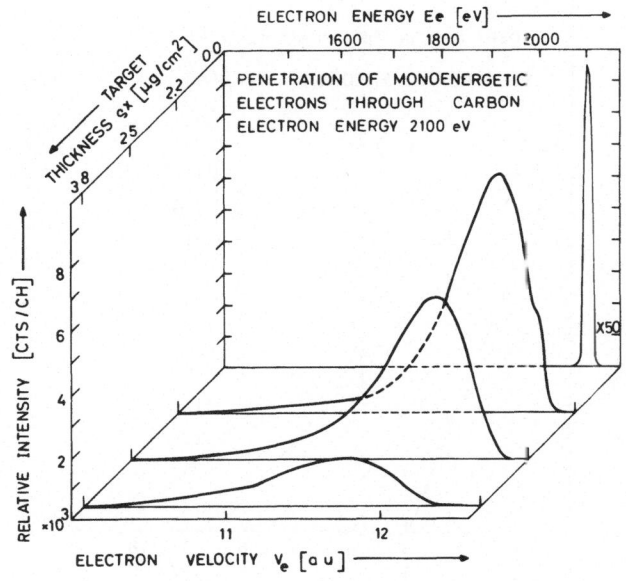

Fig.7. Electron energy distributions of monoenergetic electrons transmitted through carbon foils of different thicknesses ρx.

electron energy E_e : Typical measured electron energy distributions are shown in fig.7 as a function of the specific target thickness ρx. The characteristic feature of the spectra is an unsymmetric distribution with a broad low energy tail and a shift of the intensity maximum to lower energies with increasing ρx, which is well described by model calculations of Landau [39]. At energies E_e equal to the incident energy E_{eo} we find no significant yield of electrons without any energy loss in the target thickness range $\rho x > 2$ $\mu g/cm^2$.

The transmission T of electrons with different energies through carbon foils is well described by an exponential function T = exp ($-\mu_e \times x$) defined by μ_e (see fig.8). Fig.9 presents the strong energy dependence of μ (E_e) = $E_e^{-1.09\pm0.03}$ derived from measurements of ref. [40] and the present experiment. The numerical values of μ_e, σ_- and $<\lambda_c>$ are summarized in tab.1 .

2. We measured double differential cross sections $d^2\sigma(\theta=0)/d\Omega dE_e$ and total electron loss cross sections σ_- in carbon under single collision conditions, atomic carbon being approximated by a CH_4-gas target[41].

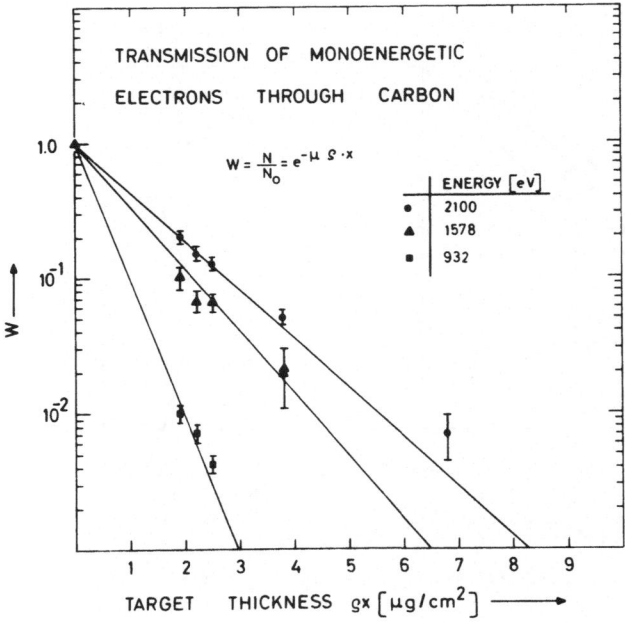

Fig.8.

Transmission of monoener-
getic electrons as function
of electron energy E_e and
specific target thickness
ρx.

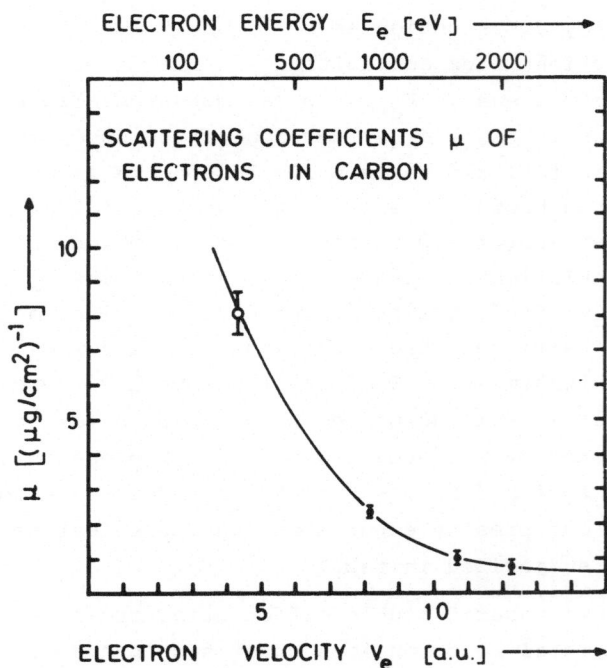

Fig.9.

Scattering coefficient
μ_e as function of electron energy E_e. The
value at $E_e = 400$ eV is
quoted from ref. [40].

Using these two sets of experimental data we calculated the ELC - fraction for He^+, H_2^+ and H_3^+ (1.7 MeV/u). As mentioned above the yield curves Y_C, for H_2^+ e.g., can be interpreted as the sum of Y_L and Y_{CS}.
It had been shown that Y_{CS} and it's angular distribution is independent on q_i (Fig.6). Let us assume that the convoy electron spectrum of isotachic H^+ accounts for the contribution of electrons represented by Y_{CS}; the convoy electron difference spectrum (CEDS) $Y = Y_C(H_2^+) - 2 \times Y_C(H^+)$ which is given by the convoy electron spectrum produced by H_2^+ minus twice the convoy electron spectrum produced by isotachic H^+, then should result in the ELC - contribution. Fig. 10 presents both the experimental CEDS (solid line) and the ELC - distribution calculated with σ_L deduced from convoy electron measurements with CH_4 - gas target (dashed line) and the ELC - distribution calculated with σ_{BS} determined from the break-up yield of H_2^+ as function of target thickness [37] (dotted line).

First, we note that the low energy tail of CEDS is well represented by the calculation using σ_L in contrast to the strong deviation - up to a factor of six - using σ_{BS}. Here, the break-up cross section σ_{BS} is not appropriate to describe the present data, because σ_{BS} accounts for both processes the   ($H_2^+ \Rightarrow 2 \times H^+ + e^-$) and the   ($H_2^+ \Rightarrow H^o + H^+$), while Y_C is only originated from the former process.

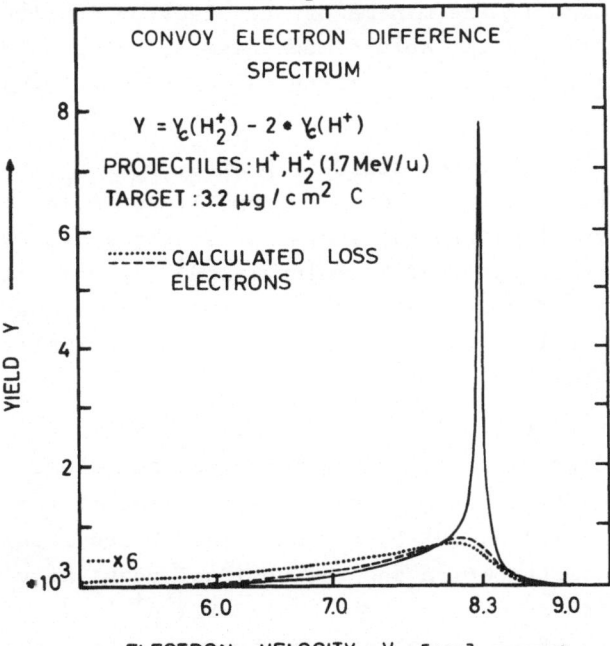

Fig. 10   Measured convoy electron difference spectrum (CEDS)
$Y = Y_C(H_2^+) - 2 Y_C(H^+)$ (solid line) and calculated ELC
distributions (dotted and dashed lines). For further
details see text.

In particular it is worth to note that the experimental data of both
atomic and molecular projectile ions show a dominant sharp residual
peak at velocities $v_e = v_p$ which is not described by our calculation.
A qualitatively similar analysis has been carried out for the $Y_C(He^+)$-
data; analogous conclusions, also concerning the residual $v_e = v_p$ - peak,
can be drawn.

The strong enhancement of the convoy electron yield with $v_e = v_p$ in
comparison to the yield of bare projectile ions suggests that these
electrons are originated from bound states of the projectile and that
they traverse through the solid with strong correlation to the ion.
Here, either a. direct or b. indirect ion - electron correlation may
contribute.

a. The direct correlation is particularly high if electrons are bound
in projectile states; it decreases rapidly with increasing quantum
numbers. This weak correlation is also valid for continuum states earlier
discussed in connection with the mean free path $<\lambda_C>$.

b. The indirect correlation is only possible in interaction of the projectile charge with the surrounding electron gas of the solid. The
responce of the medium is the damped, oscillating wake potential trai-
ling behind the ion. Here, continuum electrons may be captured into
bound states centered in the minima of the electron density fluctuation
behind the projectile. This capture mechanism is refered to as " electron
loss to wake (ELW) "   and   " electron capture to wake (ECW) " ,
respectively [42].

For atomic ions, like He^+, the residual peak at $v_e = v_p$ is only measured
in the nonequilibrium regime ($\rho x < 3$ µg/cm^2). But molecular projectile
ions show significantly different and surprising results. Fig. 11 shows
convoy electron spectra induced by molecular H_3^+ - projectiles in comparison to spectra induced by atomic H^+ - projectiles penetrating carbon
foils of different thicknesses ρx. The spectra are normalized to equal
proton numbers.

In the bottom part of fig. 11 ratio spectra $R_{SP} = Y(H_3^+) / 3 \times Y(H^+)$
are given, deduced from corresponding molecular and atomic convoy electron spectra. The left lower picture shows, in case of the nonequilibrium regime ($\rho x = 3.1$ µg/cm^2), the described broad shoulder repre-
senting the ELC - distribution and the residual peak at $v_e = v_p$. Note,
the strong yield enhancement $R = Y_C(H_3^+)/3 \times Y_C(H^+)$ ($R \leq$ eighty times
of the convoy yield of bare protons) ! The two right hand parts of
fig. 11 represent the equilibrium regime, ($\rho x = 9.5$ µg/cm^2 and
$\rho x = 20$ µg/cm^2). In contrast to results of atomic projectile ions we
find a still remaining yield enhancement R ($R \leq 2.5$ only at $v_e = v_p$).

This yield enhancement is also found by Ponce et.al. with H_2^+ - projectiles at 70 and 100 keV/u [43]. They interpret their results in terms of the effective charge q_{eff} of the correlated protons originated from the break-up of the H_2^+ - molecular projectile ion. A theoretical description of the electron capture cross section for molecular projectile ions is given by C.E. Gonzales Lepera and V.H. Ponce in this volume [44]. At the present time our preliminary investigations with heavy CO^+- and N_2^+- molecular projectile ions (0.03 MeV/u <

E_p/M_p < 0.07 MeV/u) show a similar yield enhancement in the equilibrium regime. Details will be published elsewhere [45].

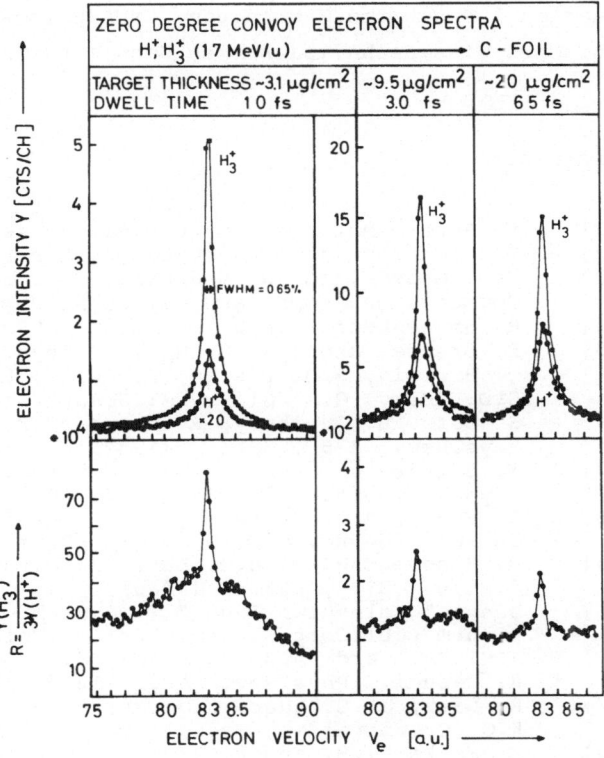

Fig. 11 Normalized convoy electron spectra of H^+ - and H_3^+ - projectiles (1.7 MeV/u) traversing carbon foils of different thicknesses ρx (upper figures) and the deduced ratio spectra $R_{sp} = Y(H_3^+)/3 \times Y(H^+)$. Note the remaining enhancement of the molecular convoy electron yield for very high dwell times.

4. Conclusion

Based on models of ECC and ELC in solids we sucessfully described the ρx - dependence of convoy electron yields Y_C of bare ions and ions with initially bound electrons and the energy distribution of ELC. The enhanced mean free paths $\langle \lambda_c \rangle$ for convoy electrons compared to these of free, isotachic electrons suggest a strong electron - ion correlation in solids, which could be possibly interpreted in the framework of Coulomb focussing and the wake model. One of the most surprising results is the observed residual peak at $v_e = v_p$. Here, we hope to elucidate the production mechanism of these electrons by the study of heavy molecular projectile ions penetrating solids of different electronic structure.

§ supported by BMFT/Bonn, NSF Physics Division and Division of International Programms/Washington, CNEA/Buenos Aires and Deutsch - Argen-
tinisches Wissenschaftsabkommen Bonn + Karlsruhe

References

[1]   N. Bohr, Mat.fys.Medd.Dan.Vid.Selsk. 18 , 1 , (1948)
[2]   K.G.Harrison and M.W. Lucas, Phys.Lett. A33 , 142 , (1970)
[3]   K.O. Groeneveld, W. Meckbach, I.A. Sellin, J. Burgdörfer,
       Comments on Atomic and Molecular Physics 4 , 187 , (1984)
[4]   R. Shakeshaft, Phys.Rev. A18 , 1930 , (1978)
[5]   F. Drepper and J.S. Briggs, J.Phys. B9 , 2063 , (1976)
[6]   M. Breinig, S.B. Elston, S. Huldt, L. Liljeby, C.R. Vane,
       S.D. Berry, G.A. Glass, M. Schauer, I.A. Sellin, G.D. Alton,
       S. Datz, S. Overbury, R. Laubert, M. Suter
       Phys.Rev. A25 , 3015 , (1982)
[7]   W. Brandt, A. Ratkowski, R.H. Ritchie,
       Phys.Rev.Lett. 33 , 1325 , (1974)
[8]   Z. Vager, B.J. Zabransky, D. Schneider, E.P. Kanter,
       Gu Yuan Zhuang, D.S. Gemmell, Phys.Rev.Lett. 48 , 592 , (1982)
[9]   D. Röschenthaler, H.D. Betz, J. Rothermel, D.H. Jakubaßa-Amundsen
       J.Phys. B16 , L233 , (1983)
[10]   see   " Molecular Ions " (J. Berkowitz, K.O. Groeneveld, ed.)
       Plenum Publ.Corp. , New York NY Vol.90 (1983)
[11]   J.M. Gaillard, A.G. de Pinho, J.C. Poizat, J. Remillieux,
       R. Saoudi, Phys.Rev. A28 , 1267 , (1983)
[12]   R. Latz, J. Schader, H.J. Frischkorn, P. Koschar, D. Hofmann,
       K.O. Groeneveld , Nucl.Instr.Meth B2 , 265 , (1984)
[13]   P. Focke, I.B. Nemirovsky, E. Gonzales Lepera, W. Meckbach,
       I.A. Sellin, K.O. Groeneveld, Nucl.Instr.Meth B2 , 235 , (1984)
[14]   M.F. Steuer, D.S. Gemmell, E.P. Kanter, E.A. Johnson,
       B.Z. Zabransky,   IEEE Transact. NS30 , 1069 , (1983)
[15]   G.J. Kumbartzki, H. Neuburger, H.P. Kohl, W. Polster,
       Nucl.Instr.Meth 194 , 29 , (1982)
[16]   H.J.Frischkorn, P. Koschar, J. Kemmler, R. Latz, J. Schader,
       K.O. Groeneveld, Nucl.Instr.Meth. B2 , 35 , (1984)
[17]   L.H. Toburen and W.E. Wilson, Phys.Rev. A19 , 2214 , (1979)
[18]   H. Folger, GSI Darmstadt, FRG ; private communication
[19]   E.F. Kennedy, D.H. Youngblood, A.E. Blaugrund,
       Phys.Rev. 158 , 897 , (1967)
[20]   R. Latz, G. Astner, H.J. Frischkorn, P. Koschar, J. Pfennig,
       J. Schader, K.O. Groeneveld, Nucl.Instr.Meth 194 , 315 , (1982)
[21]   P. Focke, W. Meckbach, C.R. Garibotti, I.B. Nemirovsky,
       Phys.Rev. A28 , 706 , (1983)
[22]   L.M. Toburen, Phys.Rev. A5 , 67 , (1971)
[23]   J.F. Ziegler, in: The Stopping and Ranges of Ions in Matter,
       vol. V (Pergamon Press, New York , 1977)
[24]   R. Schramm, D. Hofmann, P. Koschar, J. Kemmler, M. Burkhard,
       E. Rohrbach, H.J. Frischkorn, K.O. Groeneveld,
       Proc. of the 2nd Int. Conf. on " High Energy Ion-Atom Collisions "
       Debrecen/Hungary , August 1984 , (ed. D.Berenyi et.al.)
       and to be published ;   see also
       H.J. Frischkorn, K.O. Groeneveld, D. Hofmann, P. Koschar,
       R. Latz, J. Schader,   Nucl.Instr.Meth. 214 , 123 , (1983)
[25]   W. Meckbach, K.C.R. Chiu, H.H. Brongersma, Mc. Gowan, J. Wu,
       J.Phys. B10 , 3255 , (1977)
[26]   W. König, Dissertation Univ. Marburg, FRG 1979

[27]   R.W. Cranage, M.W. Lucas, J.Phys. $\underline{B9}$, 445 , (1976)
[28]   M. Rødbro and F.D. Anderson, J.Phys. $\underline{B12}$, 2883 , (1979)
[29]   A. Burgess, Mem.Roy.Astr.Soc. $\underline{69}$, 1 , (1964)
[30]   R. Shakeshaft and L. Spruch , Rev.Mod.Phys. $\underline{51}$, 369 , (1979)
[31]   M.J. Gaillard, J.C. Poizat, A. Ratkowski, J. Remillieux,
       M. Anzas, Phys.Rev. $\underline{A16}$, 2323 , (1977)
[32]   J.S. Briggs, F. Drepper, J.Phys. $\underline{B11}$, 4033 , (1978)
[33]   K. Dettmann, K.G. Harrison and M.W. Lucas,
       J.Phys. $\underline{B7}$, 269 , (1974)
[34]   H.J. Frischkorn, P. Koschar, R. Latz, J. Schader, M. Burkhard,
       D. Hofmann, K.O. Groeneveld, IEEE Transact. $\underline{NS30}$, 931 , (1983)
[35]   I.A. Sellin, S.D. Berry, M. Breinig, C. Bottcher, R. Latz,
       M. Burkhard, H. Folger, H.J. Frischkorn, K.O. Groeneveld,
       D. Hofmann, P. Koschar,
       " Electron ejection in ion-atom and ion-solid collisions "
       (K.O. Groeneveld, W. Meckbach, I.A. Sellin, ed.)
       Lecture Notes in Physics, Springer Verlag, Heidelberg (1985)
[36]   J. Remillieux, in: " Molecular Ions "
       (J. Berkowitz, K.O. Groeneveld, ed.)
       Plenum Publ.Corp., New York NY Vol. 90 (1983)
[37]   N. Cue, N.V. De Castro - Faria, M.J. Gaillard, J.C. Poizat
       and J. Remillieux, Phys.Lett. $\underline{72A}$, 104 , (1979)
[38]   R. Latz, J. Schader, H.J. Frischkorn, K.O. Groeneveld, D. Hofmann,
       P. Koschar, Z.Phys. $\underline{A304}$, 367 , (1982)
[39]   L. Landau, J.Phys. (USSR) $\underline{8}$, 201 , (1944)
[40]   I. Lindau and W.E. Spicer, J.El.Spectr. and rel.Phen. 3,408,(1974)
[41]   L.H. Toburen in Nucl.Meth. Monographs 2 , p.53
       (D. Berenyi and G. Hock, ed.)
       Sci.Publ.Company, Amsterdam 1982
[42]   Y. Yamazaki and N. Oda , Nucl.Instr.Meth. $\underline{194}$, 415 , (1982)
[43]   V.H. Ponce, C.E. Gonzalez Lepera, W. Meckbach and
       I.B. Nemirovsky, Phys.Rev.Lett. $\underline{47}$, 572 , (1981)
[44]   C.E. Gonzalez Lepera and V.H. Ponce in
       " Electron ejection in ion-atom and ion-solid collisions "
       (K.O. Groeneveld, W. Meckbach, I.A. Sellin, ed.)
       Lecture Notes in Physics, Springer Verlag, Heidelberg, (1985)
[45]   P. Koschar, J. Kemmler, M. Burkhard, D. Hofmann, R. Schramm,
       K.O.Groeneveld, M. Breinig, S. Elston, I.A. Sellin, W. Meckbach,
       Proc. of the 2nd Int.Conf. on " High Energy Ion-Atom Collisions "
       Debrecen/Hungary , August 1984 (ed. D. Berenyi et.al.)

ALIGNMENT OF HIGH RYDBERG STATES IN HYDROGEN

H. G. Berry, J. C. DeHaes,[*] D. K. Neek[†] and L. P. Somerville
Argonne National Laboratory
Argonne, IL 60439, USA

Abstract

We have measured the light yields and polarizations of the light emitted from several Balmer transitions in atomic hydrogen following beam foil excitation of protons at energies of 50 to 150 kev. The polarizations have been measured as a function of distance downbeam from the exciter foil for several transitions. The measurements indicate a very strong initial alignment which is then perturbed by surface fields out to several mm from the surface.

INTRODUCTION

Using a standard beam-foil arrangement, we have observed the fluorescence from the Balmer series in atomic hydrogen for principal quantum numbers n=4 to n=15. The initial intent of the measurements was to study the shapes of the wavefunctions of high Rydberg states after foil excitation and to check some of the earlier sug- gestions made from observations of electric-field stripped electrons from such states (1,2). Simple electron capture theory predicts populations of states proportional to n^{-3}. (From a simple density of states argument: $E \propto -n^{-2}$, leading to $dE/dn \propto 2n^{-3}$). However, some observations at lower n, and for other beam-foil excited systems, indicate different population dependences on n, which are sometimes not monotonic (3,4). Observations of the polarization of the emission indicate asymmetries in the shape of the excited state wavefunction. Typically, for lower n states, beam-foil excitation leads to wavefunctions which tend to be elongated along the beam axis - that is, the m=0 state is more populated than the m≠0 states (the beam axis being the z-axis of quantization). In figure 1 we show states where only m=0 is populated, for successively higher angular momentum values ℓ. Clearly, in general, such a population distribution corresponds to a cigar-shaped distribution along the beam axis. The opposite case, of a pancake-like wavefunction, with axis of rotational symmetry along the beam axis, corresponding to high m populations, is only rarely seen in low n-state excitation; the only examples being for hydrogen n=2 (5), and for some d-states in He I (6).

In this work, we have measured the linear polarization for the n=4 up to the n=7 Balmer lines as a function of the distance from the exciter foil. Once the atom reaches the field-free region away from the foil surface, the polarization shows quantum beats induced by the spin-orbit fine structure. These well-understood beats can be used to indicate where this field-free region begins. For example, since the

SPHERICAL HARMONICS
(m=0)

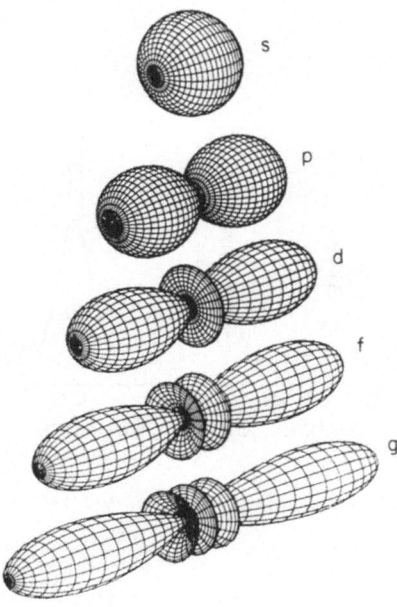

Fig. 1 Wave functions showing complete alignment along the beam-axis, with
only m=0 populated.

spin-orbit interaction is internal to the atom, the total light yield is unaffected.
This is in contrast to an external field which can change both the polarization
observed, and the light yield in both a single direction and that into all
directions. We have used this feature to draw a very surprising conclusion from our
data: the higher n states, n=5 to 7, are strongly affected by external fields out to
distances of up to 10 mm from the foil surface. Clearly, higher n states, for example
n=20 to 50, which are observed in the field-stripped electron measurements, will be
even more strongly perturbed by these external fields. We suggest an origin for this
surface electric field. However, such a field does not explain all the observations.

EXPERIMENT

Spectra were taken over the short wavelength range shown in Fig. 2 to obtain the
relative yields for high n-states n=9 to 14. Within this 100Å, the relative photon
collection efficiency remains constant to within ±10%. The grating efficiency
decreased to higher wavelengths, while the quantum efficiency of the photomultiplier
increased, both by about 4% within this range.

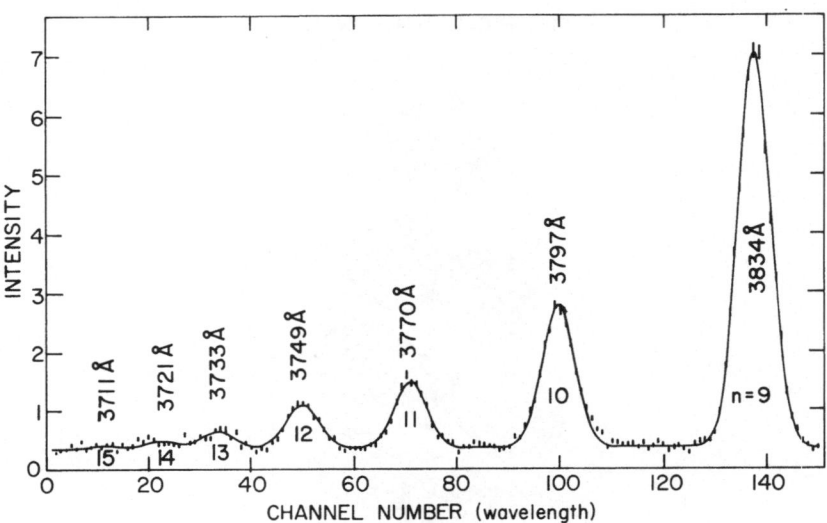

Fig. 2   Spectrum of high n Balmer transitions from 100 keV foil excited
hydrogen.

The polarization measurements were made using a standard polarimeter (7). Since only
perpendicular foils were used in these measurements, only the two linear polarization
components, I(par) and I(perp), were measured. An example of these results is shown
in Fig. 3. Four measurements of each polarization component were made at each foil
position. The foil was stepped up-beam to record the complex decay curve.

Measurements were made using H^+, H_2^+ and H_3^+ projectiles at energies between 50 and
120 keV. The molecular projectiles were dissociated by a pre-foil, generally located
about 6 mm in front of the exciter foil. The entire target chamber was surrounded
with a mumetal shield to reduce the earth's magnetic field. Without shielding, the
decay curves from the higher n-states were strongly perturbed.

RESULTS

First we fit the total light yields for the different n states. For the integrated
profiles of n=9 to 14, we obtained a power law dependence on the principal quantum
number of -(7.01 ± 0.3). In this range of n, the lifetimes except for ns states,
scale approximately as $n^{-3.1}$. Thus, the populations scale as $n^{-3.9 \pm 0.3}$. This is
somewhat faster than the scaling law predicted from the density of states model. The
next step is to analyze the quantum beat curves to find the beginning of the field-
free region for each of the decays.

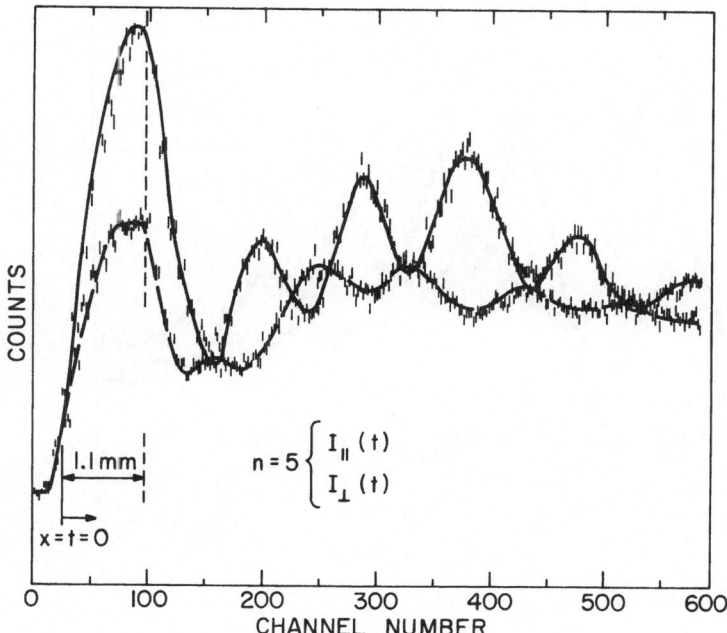

Fig. 3   Decay curves of n=2-5 observed in light polarized parallel and
perpendicular to the beam axis. Note the changes in x-scale near
the foil. Each set of 8 channels are separated by 10 times the
distance beyond channel 92 (from 0.10 to 1.0 mm).

In Fig. 4 we show the two decays for polarizations parallel and perpendicular to the
beam for the n=2-6 Balmer lines. After about 1 cm from the foil the beats are clearly
out of phase by 180 degrees, with the parallel polarization beats of double the
amplitude of the perpendicular polarization beats. This is indicative of a field-free
region, where the total light yield, I(tot)=I(par) + 2.I(perp), shows no quantum
beats, and should be a simple multi-exponential decay. We show this in Fig. 5, where
the n=4, 5, 6 and 7 total light yield decays show clearly the beginning of the field-
free region. The higher n states show stronger anomalous intensities near the foil
since their ℓ sub-states are more strongly mixed by the surface electric field. A
qualitative measure of this mixing is given by the ratio of intensities between the
first maximum and the first minimum of the total yield decay curves. This function
increases strongly with n from about 1.1 for n=4 to 5.5 for n=7.

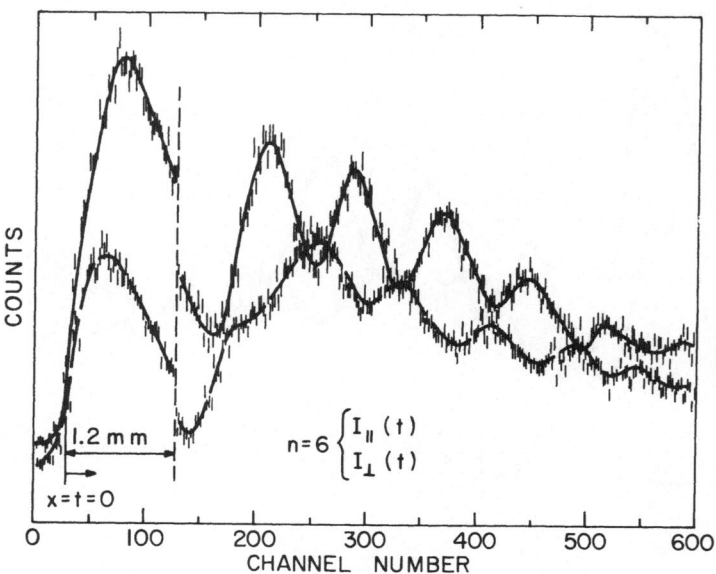

Fig. 4 Decay curves for the n=2-6 Balmer transition observed in polarizations
parallel and perpendicular to the beam axis. The x-scale change occurs at
channel 128.

The onset of the field-free region occurs at distances from 7 to 15 mm from the foil
surface. This is much larger than the few hundred angstroms expected. The field-
affected distance varies only slightly with ion-beam current or with foil thickness
between 2 and 20 $\mu g.cm^{-2}$.

We have made Fourier transforms of the polarization quantum beats in the field-free
regions for the Balmer transitions from n=4,5,6, and 7. An example of the n=4 curve
(Balmer beta) is shown in Fig. 6. Four frequencies are expected as indicated on the
energy level diagram of Fig. 7. These are found as indicated. They are observed with
similar relative intensities for the higher n states, except that the lowest frequency
beat of the $d_{3/2} - d_{5/2}$ fine structure is not measured for the n=6 and n=7 curves (its
wavelength being longer than the observation region). Including the parts of the
polarization curves within the field-affected region, of course, washes out these
observed frequencies in the Fourier transforms.

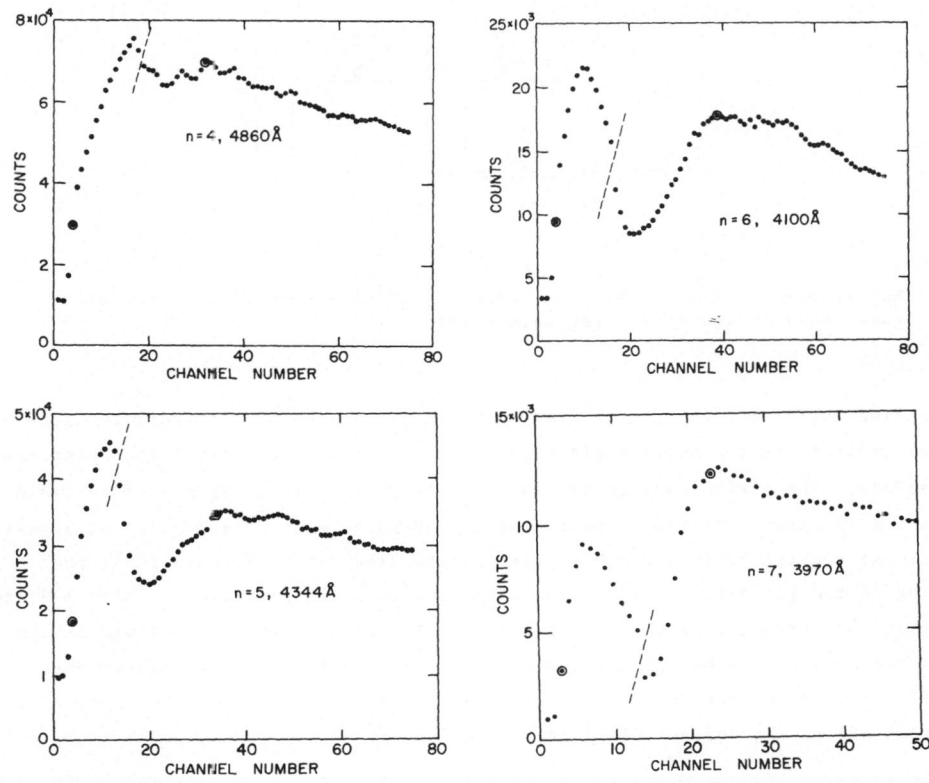

Figure 5. Total light yields, I(par)+2.I(perp), for the n=2-4, 5, 6, and 7 transitions, as functions of distance from the foil. Note the change in x-scale near the foil. The foil surface is at the first open circle. Field-free decay begins at the second open circle.

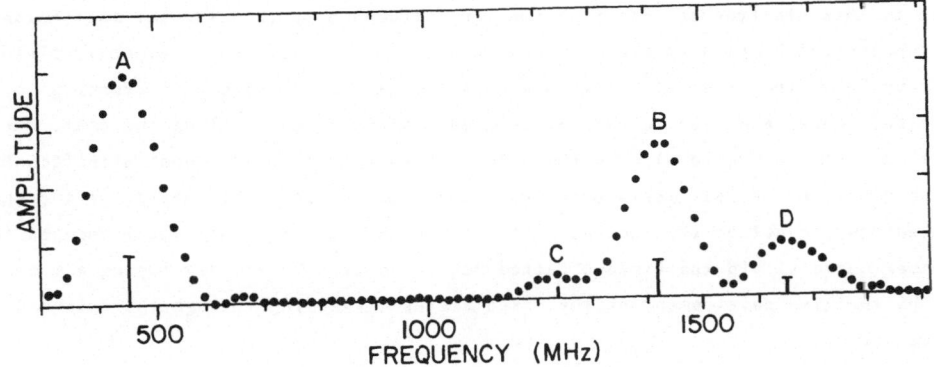

Fig. 6 Fourier transform of the n=4 polarization observed in the field-free region. The peaks A to D correspond to the fine structure splittings labelled in Fig. 7.

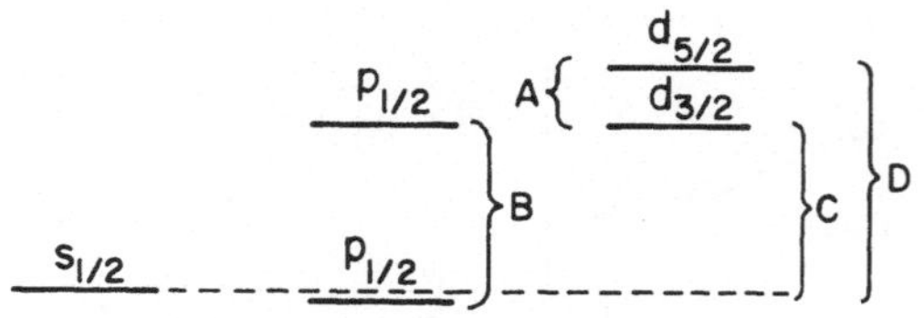

Fig. 7 Energy levels of an n-state in hydrogen indicating the fine structure
beats seen in the field-free balmer transitions.

CONCLUSIONS

We find that the light yields from high Rydberg states in beam-foil excited
hydrogen indicate strong surface electric field mixing out to about 1 cm beyond the
foil surface. The mixing is stronger for the higher n states. This surface field
reduces a very strong initial alignment at the surface, where the population appears
to be almost completely in the m=0 states (z-axis along the beam-axis). At the
beginning of the field-free region, the alignment is still positive - a high m=0 state
population, but strong mixing of the s and d states is observed. The depth of the
surface field is also shown by the unusual phase of the $P_{1/2}$ - $P_{3/2}$ quantum beat,
which is not 0 or 180 degrees relative to the foil surface, as would be expected for a
sharp turn-off of the surface field at the foil surface.

Singer et al. (8) have previously proposed an extended surface electric field to
explain previous observations of quantum beats in beam-foil excited Balmer beta. We
suggest that these measurements confirm the presence of this field and may be used to
map out its extent and other properties. Due to the finite conductivity of the thin
carbon foil, which we measure to have a diametrical resistance of a few megohms, an
electron charge is built up on the foil surface by the ion beam. The electric field
is due to this electron surface distribution surrounded by the circular foil holder at
ground potential about 3 mm from the beam center. This macroscopic electric field
will thus be of the order of a few hundred volts per cm at the center dropping to zero
at the foil edge, and also to zero at some equivalent distance along the beam. Our
results do not conform to all the characteristics of this field. Most significantly,
when we decrease the foil resistance (by increasing the foil thickness), or increase
the electron production (by increasing the ion beam current), only small changes in
the decay curve within the field-affected region occur. We are developing a model to
describe the time development of the mixed states travelling through the extended
surface field.

This research was supported by the U.S. Department of Energy, Office of Basic
Energy Sciences, under Contract W-31-109-Eng-38.

REFERENCES

*On leave from the Department of Nuclear Physics, Free University of Brussels, supported by a grant from NATO.

†Also at the University of Illinois, Chicago.

1. Z. Vager, E. Kanter, D. Schneider, and D. S. Gemmell, Phys. Rev. Lett. 50, 954 (1983).

2. E. P. Kanter, D. Schneider, Z. Vager, D. S. Gemmell, B. J. Zabransky, Gu Yuan-zhuang, P. Arcuni, P. M. Koch, D. R. Mariani, and W. Van De Water, Phys. Rev. A 29, 583 (1984).

3. S. Bashkin, H. Oona, and E. Veje, Phys. Rev. A 25, 417 (1982).

4. H. Winter, R. Zimny, A. Schirmacher, B. Becker, H. J. Andrä, and R. Fröhling, Z. f. Phys. 311, 267 (1983).

5. H. Winter, Nucl. Instr. Meths. 202, 241 (1982).

6. R. M. Schectman, R. D. Hight, S. T. Chen, L. J. Curtis, H. G. Berry, T. J. Gay, and R. DeSerio, Phys. Rev. A 22, 1591 (1980).

7. H. G. Berry, G. Gabrielse and A. E. Livingston, Appl. Opt. 16, 3200 (1977).

8. W. Singer, J. C. DeHaes and J. Carmeliet, Physica Scripta 21, 165 (1980).

Author Index

Subject Index

A

- age theory 53
- alignment 38,148
- analyzer, electrostatic
  cylindrical mirror 72
- analyzer, electrostatic
  30° parallel – plate 14
- analyzer, electrostatic
  45° parallel – plate 132
- analyzer, electrostatic
  retarding field 95
- analyzer, electrostatic
  spherical sector 76
- analyzer, 90° magnetic 132
- analyzer transmission
  function 85
- angular acceptance 106
- angular distribution,
  stationary outgoing 57
- anisotropic emission 75
- anisotropic population
  of continuum states 32
- anisotropy 62
- anisotropy parameters 41,46,48,
  108
- asymmetric collision system 17, 26
- asymmetry parameter 43
- atomic stopping power 57
- Auger electrons 94
- autoionisation lines 81
- average relative energy loss 55

B

- backward to forward ratio 62
- bare projectiles 19
- beam-foil convoy electrons 105

- beam-foil spectroscopy 116, 130 ,**148**
- Bessel function 34
- Bethe 3
- binary encounter approximation 1
- binary encounter peak 48
- Born approximation,
  first order 19,23,28,33,43,44,67,122
- second order 20,23,29,33
- Born approximation,
- **bound-bound** excitation 116
- bound state charge
  transfer 20
- bound state coherences 33
- bound state wave functions 34
- Briggs & Drepper 8
- Brinkmann & Kramers 3,4
- Brinkmann & Kramers theory 20
- bulk of solids 109

C

- capture cross section 94,116
- channeling experiments 109
- charge asymmetry 41
- charge change cross section 80
- charge equilibrium regime 140
- charge nonequilibrium regime 140
- charge transfer, bound state 20
- charge transfer into continuum 1
- classicyl trajectory Monte Carlo
  method 87
- coherence of continuum states 33
- coincidence data 109
- coincidence electron energy
  spectra 84
- collision system, asymmetric 17,26
- confluent hypergeometric function 23

P.Ring, P.Schuck

The Nuclear Many-Body Problem

1980. 171 figures. XVII, 716 pages. (Texts and Monographs in Physics). ISBN 3-540-09820-8

Contents: The Liquid Drop Model. – The Shell Model. – Rotation and Single-Particle Motion. – Nuclear Forces. – The Hartree-Fock Method. – Pairing Correlations and Superfluid Nuclei. – The Generalized Single-Particle Model (HFB Theory). – Harmonic Vibrations. – Boson Expansion Methods. – The Generator Coordinate Method. – Restoration of Broken Symmetries. – The Time Dependent Hartree-Fock Method (TDHF). – Semiclassical Methods in Nuclear Physics. – Appendices A-F. – Bibliography. – Author Index. – Subject Index.

R.Bass

Nuclear Reactions with Heavy Ions

1980. 176 figures, 31 tables. VIII, 410 pages. (Texts and Monographs in Physics). ISBN 3-540-09611-6

Contents: Introduction. – Light Scattering Systems. – Quasi-Elastic Scattering from Heavier Target Nuclei. – General Aspects of Nucleon Transfer. – Quasi-Elastic Transfer Reactions. – Deep-Inelastic Scattering and Transfer. – Complete Fusion. – Compound-Nucleus Decay. – Appendices. – Subject Index.

W.Glöckle

The Quantum Mechanical Few-Body Problem

1983. 17 figures. VIII, 197 pages. (Texts and Monographs in Physics). ISBN 3-540-12587-6

Contents: Elements of Potential Scattering Theory. – Scattering Theory for the Two-Nucleon System. – Three Interacting Particles. – Four Interacting Particles. – References. – Reviews, Monographies, and Conferences. – Subject Index.

Handbuch der Physik/ Encyclopedia of Physics

Herausgeber: **S.Flügge**
Band 31

Gruppe 6:
Röntgenstrahlen und Korpuskularstrahlen

Korpuskeln und Strahlung in Materie I Corpuscles and Radiation in Matter I

By/Von T. Åberg, G. Howat, L. Karlsson, J. A. R. Samson, H. Siegbahn, A. F. Starace

Editor/Herausgeber: **W.Mehlhorn**

1982. 268 figures. XII, 630 pages. ISBN 3-540-11313-4

Contents: Theory of Atomic Photoionization. – Atomic Photoionization. – Photoelectron Spectroscopy. – Theory of the Auger Effect. – Subject Index.

Springer-Verlag
Berlin
Heidelberg
New York
Tokyo

Lecture Notes in Physics

Selected Issues from
Lecture Notes in Mathematics